Fundamentals of Environmental Assessment

Based on the "go to" book in the field of ecological risk assessment, this shorter, principles-based, updated textbook is essential for students and new practitioners who want to understand the purposes of environmental assessments and how to achieve them. It includes environmental risks to humans as well as nonhuman populations and ecosystems, and most types of environmental assessments. Drawing upon the author's extensive experience in the field, first as a senior research staff member in the Environmental Sciences Division at Oak Ridge Laboratory and then as science advisor in the United States Environmental Protection Agency's National Center for Environmental Assessment, the book explains fundamental principles and basic techniques and illustrates them with example applications which carry through multiple chapters and make this book a practical and hands-on guide. Both the content and the style are inviting and approachable to different levels of students.

Features:

- Integrates human health and ecological assessments.
- Includes epidemiological, risk, causal, impact, and outcome assessments.
- Focuses on fundamental principles that are applicable in all nations and legal contexts.
- Employs an engaging style and draws on the author's practical experience.
- Explains fundamental concepts in short chapters, making it perfect for beginners in the field.
- Explains the challenges and rewards of a career in environmental assessment.

This book is a practical guide for senior and graduate students in environmental sciences and management, as well as new practitioners of assessment who want to understand the purposes of environmental assessments and how to achieve them.

Fundamentals of Environmental Assessment

Glenn W. Suter II

CRC Press is an imprint of the
Taylor & Francis Group, an **informa** business

Designed cover image: © iStock

First edition published 2023
by CRC Press
6000 Broken Sound Parkway NW, Suite 300, Boca Raton, FL 33487-2742

and by CRC Press
4 Park Square, Milton Park, Abingdon, Oxon, OX14 4RN

CRC Press is an imprint of Taylor & Francis Group, LLC

Library of Congress Cataloging-in-Publication Data
Names: Suter, Glenn W., II, author.
Title: Fundamentals of environmental assessment / Glenn W. Suter, II.
Description: First edition. | Boca Raton : CRC Press, [2023] |
Includes bibliographical references and index. | Summary: "Based on the "go to"
book in the field of ecological risk assessment, this shorter,
principles-based, updated textbook is essential for students and new
practitioners who want to understand the purposes of environmental
assessments and how to achieve them. It includes environmental risks to
humans as well as nonhuman populations and ecosystems. Drawing upon the
author's extensive experience in the field, the book explains
fundamental principles and basic techniques and illustrates them with
example applications which carry through multiple chapters and make this
book a practical and hands-on guide. Both the content and the style are
inviting and approachable to different levels of students"-- Provided by publisher.
Identifiers: LCCN 2022056429 (print) | LCCN 2022056430 (ebook) |
ISBN 9780367705923 (hbk) | ISBN 9780367741563 (pbk) | ISBN 9781003156307 (ebk)
Subjects: LCSH: Environmental impact analysis. | Environmental risk
assessment. | Health risk assessment.
Classification: LCC TD194.6 .S88 2023 (print) | LCC TD194.6 (ebook) |
DDC 333.71/4—dc23/eng/20230109
LC record available at https://lccn.loc.gov/2022056429
LC ebook record available at https://lccn.loc.gov/2022056430

ISBN: 978-0-367-70592-3 (hbk)
ISBN: 978-0-367-74156-3 (pbk)
ISBN: 978-1-003-15630-7 (ebk)

DOI: 10.1201/9781003156307

Typeset in Palatino
by codeMantra

Contents

Part III Methods for Environmental Assessment

Part IV Supporting Material

Preface: What Kind of Book Is This?

The first edition of *Ecological Risk Assessment*, completed in 1992, was the first text in its field. The second edition, completed in 2006, was longer and dove more deeply into the practice of ERA. By then, the field was 20 years old, and that experience went into the second edition. I was proud of it, it sold reasonably well, and I received compliments. However, there were clear signs that it did not suit everyone. The first review in Amazon was "Too technical for being called a textbook." An academic colleague dismissed it as "encyclopedic." A professional colleague described it as "dense." A good friend and colleague read little of her gift copy because it was "not engaging." Finally, I was told that it was too long for a text and that students these days are put off by large books. To me, a risk assessment book should be technical, it should cover the subject completely (encyclopedic), and technical writing should present the information efficiently (dense). However, I had to admit that my idea of what that book should be was not universally shared.

This book is intended to address those issues, update the material, and appeal to more readers.

1. It is shorter.
2. It is more focused on ensuring that the reader understands the fundamental concepts of environmental assessment and not so much on the specifics of models, tests, chemicals, etc.
3. It is focused primarily on risk assessment but includes other types of environmental assessment. Risk assessment is most common, but students and practitioners are likely to encounter other types.
4. It is primarily ecological but includes some human health to encourage integrated assessments. The basic principles and processes are the same.
5. Its style is more personal. I have tried to avoid the typical "voice of God" style found in scientific texts and explicitly addressed the reader and presented my experience and opinions.
6. Its new name expresses these changes.

Environmental assessment has matured. Government agencies (USEPA, EFSA, ECHA, etc.) have produced detailed and expansive guidance documents. Government and industry have both produced many exemplary assessments of chemicals, effluents, projects, and contaminated sites. Once the basic principles are understood, an assessor can easily find guidance, information, and model assessments appropriate to their assessment.

My primary audience is environmental assessment beginners. These may be students or scientists who, like me in 1975, have been trained in a related field and are diverted into assessment. I hope that current practitioners will also find it worth their while. It makes points that previous assessment books did not.

The approach to environmental assessment that I present is largely mainstream, in part because I did a lot of the digging of that channel. I also wasted time digging channels that the mainstream never flowed through. My four-decade accumulation of experience has made me a pragmatist and made this book pragmatic. I emphasize assessment practices that are effective. At ORNL, I worked for responsible parties (organizations that are responsible for contamination and disturbance), the Atomic Energy Commission and Department of Energy. Then I moved to a regulatory agency, the U.S. Environmental Protection Agency. I know what is expected in both contexts and how useful and high-quality science can be done in both contexts. Many of my publications are frameworks, methods, and guidance, so that you can effectively solve environmental problems. Most of the framework and methods are those of government agencies. If you are working in industry, the processes are likely to be much less complex than in agencies. For example, you are unlikely to be required to perform public briefing or public review, comment, and responses.

Because this book is about principles and not how-to, it contains few citations relative to my prior books. If, for example, you want to know how to derive a concentration–response model, it is best to go to the web and find the latest agency guidance and recent exemplary assessments. Methods go out-of-date faster than principles. The relatively few citations support facts that are not common or readily available knowledge, exemplary or interesting cases, policies, and guidance documents that provide a starting point for your own work. Many citations reference U.S. Environmental Protection Agency documents. They are relevant to U.S. readers and are often helpful to assessors elsewhere. With the help of the web, you can find equivalent documents for the European Union and other governments with well-developed environmental protection programs.

Examples in this book rely heavily on a few of the assessments in which I participated. They include the National Acid Precipitation Assessment (Baker et al. 1990), the assessment of the cause of the decline in San Joaquin kit foxes on the Elk Hills Naval Petroleum Reserve (Suter and O'Farrell 2009), the field-based aquatic life benchmark for conductivity (USEPA 2011b), multiple Superfund assessments of the Oak Ridge Reservation (no longer available), and the Bristol Bay watershed-scale risk assessment of the proposed Pebble Mine (USEPAR10 2014). This is not because I am self-centered or because I am dismissing other people's work. Rather, it is because I know more about the processes that generated those products than is included in the other people's worthy assessments.

Acknowledgments

This is my eighth book and swan song as a book author, so acknowledgments are important. First, thanks to my colleagues. Lone geniuses are rare, and I certainly am not one. My accomplishments have involved wonderful collaborators, particularly Susan Cormier, Larry Barnthouse, Sue Norton, Rebecca Efroymson, Brad Sample, Bob O'Neill, Leo Posthuma, Kate Schofield, and Lucina Lizarraga. Second, great thanks to my wife of 54 years who has saved me from the pitiful and lonely life that would otherwise have been the fate of a nerdy introverted guy.

Author

Glenn W. Suter II is a retired environmental assessor. During his 20 years at the U.S. Environmental Protection Agency, he served as the Science Advisor to the Director of the National Center for Environmental Assessment in Cincinnati and as Chairman of the Risk Assessment Forum's Ecological Oversight Committee. Before that, he worked for 23 years in the Environmental Sciences Division of Oak Ridge National Laboratory. He has authored more than 300 publications including four authored books and four edited books. Among his recognitions, he received the SETAC Founder's Award and the USEPA's Level One Scientific Achievement Award, and gold medal for exceptional service. He is a Fellow of the AAAS and SETAC. He has worked on the development of toxicity test protocols; environmental impact assessment for nuclear, geothermal, and coal energy technologies; the National Acid Precipitation Assessment; recovery of an endangered species; and ecological risk assessments of contaminated sites and of the mining of copper and coal. He participated in the development of guidance for ecological risk assessment, causal assessment, weight of evidence, field-based water quality criteria, assessment endpoints, ecosystem services, integrated risk assessment, and other topics. After this book, he intends to spend more time clearing invasive species and otherwise restoring his bit of mixed deciduous forest.

Prologue: Glenn's Dead Ends and a Few Good Ideas

I had a role in developing some of the ideas in this book, in collaboration with others. If you are in on the ground floor of a new practice like environmental assessment, you inevitably invent things and set precedents. It is tempting to leave out those ideas that were unsuccessful, but that would be misleading. Many ideas that I contributed to since 1975 seemed great at the time but turned out to be dead ends. That is, they are seldom if ever used, because they are impractical, do not provide useful results, or are just wrong. However, some of them are still advocated by other assessors. The dead ends include the following:

Risk assessments must be probabilistic: When our group at ORNL was asked by the USEPA to develop a risk assessment practice for the environment equivalent to human health risk assessment, we decided that it would be distinguished from other ecological assessments by being probabilistic (Barnthouse and Suter 1986; Barnthouse et al. 1982). The belief that risk assessments must be probabilistic is still commonly held, but in practice it is not necessary, is seldom useful in decision making, and may be misleading (Chapter 18). Probabilistic assessment is still advocated, but very seldom by environmental regulators or managers.

Prefer complex methods: It is tempting to assume that since nature is complex, complex tests and models are best (Chapter 19). In the early 1980s, a group of us at ORNL developed and evaluated complex multispecies tests for the USEPA Office of Toxic Substances (Garten, Suter, and Blaylock 1985; Suter 1983; Hammons 1981). Tests that we developed were adopted by the Office of Toxic Substances but not used by manufacturers. For a while, the USEPA pesticides office requested mesocosm tests for upper tier assessments, but they were dropped partly due to expense, but largely due to difficulty in interpretation. That is, of the many things happening in the tests, what constitutes adaptation versus an actionable adverse outcome? Similarly, the first ecological simulation model developed specifically for ecological risk assessment was our lake ecosystem model which included exposure and toxic effects on all trophic groups (O'Neill et al. 1982). The USEPA later developed a similar but much improved model (AQUATOX, https://www.epa.gov/ceam/aquatox) which has served to gain understanding but not inform decisions. For various reasons, decision-makers have been reluctant to use complex tests and models for setting criteria, deriving clean-up levels, determining the cause of an impaired water, etc. (Chapter 19). Complex assessment tools may be useful in specific cases but are not inherently preferable.

Prefer new methods, especially your own: We as scientists are taught that science progresses, so the latest methods and results are best. Plus, if you develop your own novel methods, you can gain publications and recognition. However, novelty for its own sake in a legal context is likely to be perceived as arbitrary and capricious and will not necessarily provide better information. New methods should be needed because some purpose is not being served by existing methods, but such justifications are seldom provided. Novelty is probably the most tempting dead end to go down.

Presume causation: We are all taught that association is not causation, but we assume it anyway. If an effect occurs where exposure occurs, particularly if the correlation is strong, we tend to go with it. I, like most assessors, treated causal assessments as if they were risk assessments until Susan Cormier recruited me to work with her and Sue Norton on guidelines for determining the causes of biological impairment (USEPA 2000b) (Chapter 12). Causal assessment is an important and challenging but necessary type of assessment.

Integrate human health and ecological assessments: I advocated integration of risk assessments for Superfund on the Oak Ridge Reservation and in the literature (Suter 1997). After joining the USEPA, I participated in a World Health Organization project to develop an integrated framework for human health and ecological assessment (Suter et al. 2003; WHO 2001). Although integrated assessment has been adopted in Europe, it has not become USEPA policy (Vermier et al. 2007). My attempt to develop common guidelines for weight of evidence in the USEPA was rebuffed by the human health assessors. Integration, to the extent that it occurs in the USEPA, is informal and case specific. The human health and ecological stove pipes are firmly embedded. However, I still believe that this is a noble cause, and even if there is no formal integration, you can communicate across the divide (Chapter 14).

Assessors should avoid normative considerations: This is the mandate of some purists who say that scientists should stick to science and avoid value judgments. However, useful assessments are inherently normative because they implement laws that are codified norms (Chapter 14). Decision-makers want either a standard decision criterion or, if unfamiliar methods are used that produce nonstandard results, they want a normative interpretation that relates the results to regulations or precedents. In one case decades ago, a regulatory manager asked me how many earthworms are enough? I said it depends on his goal which he should identify, and at that point, he dismissed me. Because I had failed to share the burden of relating science to norms, I was useless.

Successful ideas are those that have been adopted by assessors and that influence environmental management. I can claim early contributions to ecological risk assessment, the use of frameworks for assessment processes, planning processes that are more than hazard identification, weight of evidence, and species sensitivity distributions. These ideas were not obviously better than the dead ends. One can ask the regulators who use assessment tools or results what is needed, but users do not necessarily know what will be useful until they use it. Ultimately, you must consult decision-makers and your assessor colleagues before proceeding, and monitor their responses for signs of adoption or rejection.

Part I

Types and Components of Environmental Assessment

This book addresses all types of environmental assessment. It primarily explains ecological assessment but relates it to human health assessment which is simpler. It begins with the basic types: assessments of conditions, assessments of causes, predictive assessments, and assessments of outcomes. It then explains what assessors do and how they relate to others involved in environmental decision making. The rest of this part presents frameworks for environmental assessment, and the major planning, analytical, synthesis, and communication steps. It ends with a summary of pollution control methods so that assessors know what sorts of management options might be informed by their assessments.

DOI: 10.1201/9781003156307-1

1

Types of Environmental Assessment

Know the why of what you are doing, and the how will follow.
—*Fredrich Nietzsche,* The Twilight of the Idols

If you are reading this, good for you. Either you are an environmental assessor, or you are preparing to become one. You have chosen to make a difference in the environment. Environmental assessment is a process of generating and presenting scientific information to inform environmental management decisions. By turning data into actionable information, you address hazards to humans and to nonhuman organisms, populations, communities, and ecosystems from agents in the environment.

Assessors must be aware of the decision context and the type of assessment that it requires. It is helpful to organize the types of assessments in a 2×2 matrix (Figure 1.1). The left column includes assessments that detect problems, either before (condition) or after (outcome) a management or regulatory action. The right column includes assessments that resolve problems, either by determining the cause of adverse conditions (causal) or by predicting the results of alternative actions (predictive). The top row is environmental epidemiology which identifies human and ecological problems and their causes. The bottom row is environmental management which determines what action to take and whether it worked.

For a complete environmental assessment cycle, you would begin with a condition assessment to determine whether there is an environmental impairment, and then perform a causal assessment to identify the cause of any identified impairments. If the cause is within the purview of your organization, you move on to a risk assessment or other predictive assessment to inform the choice of an appropriate action. Finally, you perform an outcome assessment to assure that the action succeeded or to determine the nature of any failure. You will not often follow this full cycle, because it relates only to existing hazards and because a full cycle is likely to involve different teams. You can begin the process with any of the four types of assessment and end it with any assessment that resolves the issue. Most environmental assessments begin with a predictive assessment of a future hazard—new chemical, a proposed project, or potential remediation.

DOI: 10.1201/9781003156307-2

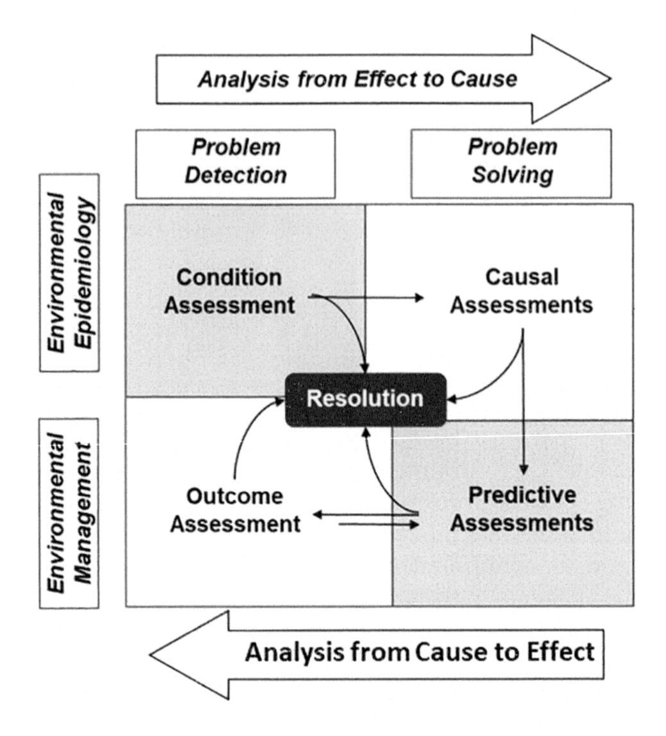

FIGURE 1.1
A process framework for the types of environmental assessments and relationships among them. (Modified from Cormier and Suter 2008.)

1.1 Condition Assessments

Condition assessments include all assessments that determine the acceptability of observed environmental conditions. For example, does the water exceed any ambient water quality criteria, is the ambient water or effluent toxic, is an endangered species declining, is the biotic community impaired, how fast is the global mean temperature rising, or is the childhood leukemia rate elevated? Condition assessments are a basic tool of regulatory enforcement. If a standard or criterion is exceeded with the specified frequency and duration, the regulatory agency will levy the fine or take other action. The impetus for a condition assessment may be a regulatory requirement for regular monitoring and reporting. Condition assessments may also be prompted by reports of apparent problems such as foaming or colored water, bad smells, deformed fish, and human illnesses.

A condition assessment based on air or water quality monitoring is a relatively straightforward application of sampling statistics and analytical chemistry. In a regulatory context, sampling and chemical analysis methods and frequencies are specified, and statistical analyses are usually minimal. These include determination of compliance with emission permits or with ambient air and water quality standards.

An important type of condition assessment in the U.S. is wetland delineation. It determines whether a location is a wetland that is afforded protection from excavation or filling. This assessment is so common that the U.S. Army Corps of Engineers has developed an automated wetland determination system that considers hydrology, soils, and biota (Schulz and Berkowitz 2017).

A more complex example of condition assessments is provided by determination of whether nominally contaminated sites should be added to the Superfund National Priorities List (NPL). This process may involve three tiers of assessment conducted by a state or the USEPA. Quick initial screens are performed, because some contaminated sites are actually clean (someone misremembered where waste was dumped), or the waste is innocuous (it was nontoxic or just clean fill). A second tier is more focused, but it still relies on existing data, records, and memory. The final tier employs the Hazard Ranking System (HRS) toolbox (https://www.epa.gov/superfund/hrs-toolbox) that provides guidance and tools such as soil screening levels to derive an HRS score. If the site HRS score is high enough, the site is added to the NPL where it waits for the actual risk assessment (remedial investigation) to determine the need for remediation.

The condition of human populations is determined by basic demographic and public health data. This practice began in 1662 when John Graunt invented demographics and used it to document the consequences of the Black Death. Modern epidemiological assessments use population data to determine the frequency of a particular health impairment in a population and the likelihood that it is due to chance. Such studies may be prompted by reports of elevated effects in a local population termed a cluster (Abrams et al. 2013). Highly publicized investigations of clusters of cancer or birth defects in contaminated communities such as Times Beach, Missouri; Love Canal, New York; Woburn, Massachusetts; and Toms River, New Jersey have raised public concerns about pollution. Human condition assessments may also be driven by known contamination in the absence of known or alleged effects. Such synoptic studies may be conducted at contaminated sites by examining public health records or interviewing potentially exposed residents. However, in the U.S. Superfund program, few sites have enough exposed people to confidently detect effects, so health protection has relied on a risk assessment approach. In contrast, the major air pollutants cause extensive exposures and health effects that have been documented by numerous epidemiological studies.

Ecological condition assessments are less straightforward than those for physical–chemical or human health conditions because of the plethora of potential measures of impairment. The USEPA Office of Research and Development (ORD) created the Environmental Monitoring and Assessment Program (EMAP) to develop and demonstrate methods for assessing status and trends in ecological resources. A probability sampling strategy and standard sampling methods were developed for consistency across scales and regions. The problem of defining biological impairment was addressed by deriving multimetric indices. EMAP's monitoring and assessment program was picked up by the Office of Water in collaboration with states and tribes as the periodic National Aquatic Resource Surveys (NARS). In addition, some states and tribes continue to use versions of EMAP methods to identify biologically impaired waters along with chemical contaminant monitoring to comply with the U.S. Clean Water Act. The European Water Framework Directive has an equivalent but more complete program for assessing ecological and chemical status based on surveillance monitoring (European Union 2000). European water bodies are rated high, good, moderate, poor, or bad biological quality based on four groups: phytoplankton, macrophytes and phytobenthos, benthic invertebrate fauna, and fish fauna.

The key question in such condition assessments is, is the apparent impairment real? In some cases, such as the extinction of peregrine falcons in much of their range or any large fish kill, the reality of the impairment is manifest. Rafts of dead fish floating down the stream are not normal. One needs only to document it and move on to assessing causation. In many other cases, including nearly all of the human health cases, it is necessary to determine that the cluster of cases, difference in abundance, or other apparent impairments are unlikely to be due to random variance. It is no longer considered appropriate to statistically test a no-effect hypothesis, although that is still done (Chapter 24). Rather, one should determine the magnitude to the apparent effect relative to a reference such as public health records or reference ecosystems (Chapter 24).

Various factors may influence the decision that the apparent impairment is sufficiently likely to be anthropogenically caused rather than naturally random. For example, in the San Joaquin kit fox case, the species was listed as endangered, and the U.S. Department of Energy was faced with shutdown of the Elk Hills Naval Petroleum Reserve, so arguing that the decline was random variance was not an option. A causal assessment was begun in the hope of absolving oil development or identifying a readily remediated contaminant.

In some cases, the reality of the phenomenon may remain up for grabs. It was never clear whether colony collapse disorder of honeybees was a real disorder or just a rare behavior that became relatively common in 2006–2007. In any case, reported occurrences of colony collapse disorder—as opposed to the general decline in honeybees due to disease, parasites, pesticides, and habitat quality, which clearly is occurring—ended before anything was resolved.

If the impairment is found to be real, the process may follow different paths. It may be deemed real but irrelevant or unimportant, and the issue is resolved. It may be real, and the cause may be obvious, so you move on to a predictive assessment of alternative actions. Or it may be real, but the cause is unknown or uncertain, so a causal assessment is performed. Causal assessments may be an extension of the condition assessment in that finding a cause should increase confidence in the reality of the effect. If the reality of an impairment is uncertain (e.g., a cancer cluster is small), identifying a cause (e.g., high levels of a carcinogen in the drinking water) can be evidence against random variance as an explanation. Similarly, skepticism concerning the reality of colony collapse disorder in honeybees was supported by the failure to identify a cause.

An uncommon type of ecological condition assessment begins with a known cause and effects, and determines the significance of effects in the field. For example, lead poisoning is known to occur in raptorial and scavenging birds, but the magnitude of effects has been uncertain. A study of bald and golden eagles determined that tissue concentrations and frequencies of acute and chronic effects of lead were higher than expected and sufficient to suppress population growth rates (Slabe et al. 2022). Because the agent of concern is known, this type of study might be considered a risk assessment, but I included it here because the approach is ecoepidemiological. Studies of this type are more common in human health epidemiology.

1.2 Causal Assessments

We environmental scientists have been concerned from the beginning with causes of impairments. Environmental causes of outbreaks of human illness have been a concern of epidemiologists since John Snow showed that cholera is caused by a water-borne pathogen. All epidemiology texts contain discussions of causation, but they differ remarkably in their view of appropriate assessment methods and criteria (Chapter 11).

Ecological causal assessment, as a recognized type of assessment, began in 1998 when the USEPA Office of Water decided that they needed a method for determining the causes of biological impairments. Some states had adapted the techniques developed by EMAP for biological status and trends assessment to identify biologically impaired waters which were then listed. However, without knowing the causes of those impairments, the states could not correct them as required by the Clean Water Act. The Office of Research and Development developed guidance for causal assessment (USEPA 2000b) and an expert system, the Causal Analysis, Diagnosis, Decision Information System (http://www.epa.gov/caddis/) to help assessors. Many assessments make causal conclusions for observed impairments, but they are usually methodologically informal (Chapter 11).

Classic ecoepidemiology cases, such as the extinction of peregrine falcons in much of the U.S. and Europe, forest dying in the 1980s (Waldsterben in Germany), frog deformities in the U.S., and oyster deformities in France, involved larger and clearer effects than human health cases. My personal experience includes the decline in San Joaquin kit foxes on the Elk Hills (Suter and O'Farrell. 2009) and the decline in taxa richness of aquatic insects in Central Appalachia (USEPA 2011b; Cormier, Suter, et al. 2013).

Causal assessments may determine two causes of observed impairments: the proximate cause such as a chemical exposure and the source such as an effluent, waste site, or pesticide application. They may proceed in any of three ways:

1. You may specify an effluent, waste, application, or other source as the source of concern, because it is a target for regulation or it is a source for which you are responsible. The causal assessment must determine whether it is the cause of an observed impairment and what agents from that source are proximate causes. Your causal assessment must determine a pathway from source to exposure for a constituent and then determine its sufficiency as a proximate cause. For example, in the Elk Hills kit fox case, oil field wastes were suspected of causing the kit fox decline and that source was the only cause of concern for the managers. Potential pathways to kit fox exposure to wastes were identified, but exposures were insufficient. By eliminating all waste-derived proximate causes, the wastes were eliminated as causal sources (Chapter 24).
2. If you know a proximate cause, the causal assessment must determine its source by linking source to exposure. For example, the toxic cyanobacterial blooms in Lake Erie that contaminated the drinking water for Toledo, Ohio, were believed to be caused by elevated phosphorous concentrations. The question for the causal assessment was to what extent did the phosphorus come from agriculture, sewage, or some other source? Such assessments are used to determine waste load allocations. Similarly, the acidification of lakes in the Adirondack Mountains of New York was caused by acid deposition, but much monitoring and modeling effort in the NAPAP assessment was required to determine that Ohio Valley coal-fired power plants were the primary source of that proximate cause. Some sources can be identified by techniques such as identifying isotopes, ions, relative concentrations of constituents of chemical mixtures that are characteristic of a particular source.
3. Some of the most interesting causal assessments begin with an impairment with unknown sources, unknown proximate causes, and unspecified sources of interest. As a causal assessor, you must form plausible hypotheses, and for each, infer both the proximate cause and its source. For example, the cause of the precipitous decline and extinction of peregrine falcons in most of the U.S. and much of Europe was unknown, and hypotheses were all over the map: loss of prey, shooting, egg collecting, disease, and contaminants. Similarly,

the cancer clusters in Woburn, Massachusetts, and Toms River, New Jersey, were suspected to be due to pollution, but initially neither potentially causal chemicals nor their sources were known.

Since neither correlation nor any other measure of association can establish causation, all of these causal assessments should be assessed by weighing the evidence (Chapters 24, 25).

1.3 Predictive Assessments

Prediction of the environmental consequences of actions is the most common type of environmental assessment. As an environmental assessor, you can expect to predict the consequences of projects, chemical uses, effluents, standards, species introductions, remedial actions, or other decisions with environmental consequences.

1.3.1 Impact Assessment

The U.S. National Environmental Policy Act of 1969 (NEPA) requires that the environmental impacts of major federal actions be assessed. Since then, many nations have enacted environmental protection laws modeled on NEPA that require impact assessments. Impact assessments predict the potential effects of projects such as highways, dams, forest harvesting, and power plants that may be carried out, funded, or permitted by the government. More than other assessments, NEPA assessments are focused on public disclosure.

The U.S. impact assessment process is prescribed by Council on Environmental Quality regulations. There are three assessment steps. First, the project may be given a categorical exclusion because it belongs to a category of actions that were found to not cause significant effects. Ironically, the licensing of the Deepwater Horizon well, which caused the biggest oil spill on record, had been given a categorical exclusion. Second, a relatively brief environmental impact assessment (EIA) may determine that the action will not have a significant effect on the human environment and the agency will issue a finding of no significant impact (a FONSI). Third, if there is no categorical exclusion or FONSI, an environmental impact statement (EIS) is prepared. By law, an EIS should determine and present the environmental impacts of the proposed action, particularly unavoidable adverse impacts. It should also present the reasonable alternatives, the relationship between short-term uses and long-term productivity, and any irreversible and irretrievable commitment of resources.

Impact assessments vary greatly among agencies and types of actions. They tend to be more descriptive than other assessments because of the requirement that they be full disclosure documents. For example, they may have lists of flora and fauna of all biotic community types in the path of a proposed highway and the area of each type that would be destroyed. However, some EISs are highly analytical. For example, assessments of projects that involve the taking of endangered species or loss of habitat have led to dueling population models from land management agencies (e.g., the U.S. Forest Service) and species trustee agencies (e.g., the U.S. Fish and Wildlife Service).

Impact assessments have the obvious benefit of introducing environmental considerations in contexts where they would otherwise be ignored or treated superficially. However, they have been criticized for being expensive and holding up progress for months to years without significant environmental benefit. The lack of apparent benefits occurs in the U.S., at least in part because impact assessments are often prepared by an agency that wants to carry out a project and sees the impact assessment as just a legal hurdle to clear. Environmental advocacy groups argue that impact assessments just whitewash pre-determined decisions and project advocates argue that they are just an opportunity for environmentalists to delay a project with lawsuits. I know from my impact assessment experience with proposed nuclear power plants, "clean" coal plants, and geothermal projects that impact assessments can result in major changes such as adding cooling towers, rerouting power lines, and even canceling projects. They are more effective than their reputation implies because many of the benefits occur behind the scenes and before the assessment is completed.

The mixed narratives about impact assessment have led to modifications of processes to make them more efficient and effective. The CEQ 1973 guidance was revised in 1978 to make EIAs more analytical and also more understandable. Some efforts have been made to make EIAs more rigorous by applying the ecological risk assessment approach (Efroymson et al. 2001; Suter et al. 2002). The new Canadian Environmental Assessment Act of 2012 focuses EIAs on major projects and sets a 365-day limit for completion. It also makes the assessments more substantive by specifying that their conclusions should include conditions on the project that may apply to a permit or may be enforced because they are in the assessment. Similarly, the EU amended the EIA Directive in 2014 to make the procedures simpler with prescribed timeframes, elevated quality, and more comprehensible presentation. The proper balance between efficiency and full disclosure of impacts is still being sought.

Impact assessments under NEPA include programmatic assessments, which, as the name implies, address programs rather than individual projects. They typically include many constituent projects, occurring over a long timeframe, and a large geographic area. They may be based on implementation of a new policy with new objectives. Examples include assessing a water management plan for a river rather than a single dam or a new regional forest management plan rather than a single timber sale.

In many nations, the term strategic environmental assessment is used often with an emphasis on very large scales and multiple resources and objectives.

1.3.2 Risk Assessment

Environmental risk assessments (ERAs), like impact assessments, predict effects on the environment. ERAs differ in that in the U.S. they are not required by law, are not focused on projects, and are less focused on public disclosure. Rather, they have been adopted by environmental agencies as tools for informed decision making. This decision focus in risk assessment began in the 17th century with the calculation of insurance premiums and other uncertain financial predictions (Bernstein 1996). For example, if you financed a voyage in the 17th-century spice trade, you asked what are the expected profits and the chance that the ship will return, because you were putting your money at risk. This concept of risk assessment as a process for informing decision making under uncertainty carried over into engineering and regulation of pharmaceuticals, foods, devices, effluents, chemicals, and wastes. They are also used to assess nonchemical agents including suspended particles, thermal effluents, radionuclides, introduced species, genetically modified organisms, and other agents that are thought to pose a hazard.

On the industry side, risk assessments are performed to inform product stewardship and ensure that a product will not damage human health or the environment. One variant is product life cycle assessment which attempts to address all steps "from cradle to grave" including raw material extraction and processing, product production, distribution, use, and disposal. Also, most manufacturers or effluent emitters may perform their own risk assessments to anticipate the regulatory agency's findings. As a result, they may modify or even withdraw a product or increase treatment of a waste. Finally, product defense consultants perform risk assessments for industry to influence regulators and their reviewer panels or to provide a basis for expert testimony in court.

Risk assessments tend to be structured and analytical. The structure appears as process diagrams called frameworks (Chapter 3). The analyses include transport, fate, and exposure processes (Chapter 6); hazards and exposure–response relationships (Chapter 7); and characterization of effects, confidence, and uncertainty (Chapter 8).

1.3.3 Benchmark Assessment

Comparison of a pollutant concentration in ambient media to a legal standard is the type of risk assessment that has done the most to reduce environmental pollution. Its success has depended on the development of benchmark values variously called criteria, standards, guidelines, reference doses, limit values, and bright lines. Because the assessments that derive these values do not have a generally accepted name, we now call them benchmark assessments to encompass all of the variants (Chapter 22). They are a type of predictive assessment in that the assessors are predicting that if the value that they derive is not exceeded, unacceptable effects will not occur (Suter and Cormier 2008). The degree to which they are precautionary depends on their intended use. Screening benchmarks are deliberately precautionary because, if they are not exceeded, the data gathering and assessment will end early. If they are an enforceable standard, they will be more balanced between protection and avoidance of unnecessary restrictions.

Like other assessments, benchmark assessments depend on some sort of relationship between exposure and response. However, unlike exposure–response relationships in risk assessments which are solved for an exposure level, in the derivation of benchmarks, the relationships are solved for a level of effect (Figure 1.2). Two methods have been regularly used to derive the exposure–response relationships for benchmarks. First, historically and currently, both human health and ecological assessors have chosen a benchmark value from available results of toxicity tests. Then, if that value is not sufficiently relevant or reliable, it is divided by one or more assessment factors. Second, when sufficient ecotoxicological test data are available, a species sensitivity distribution (SSD) is used as a community exposure–response relationship (Chapter 27). The benchmark value is a low centile of the distribution, usually the 5th, which implies that 95% of species will be protected. Other possible methods have been proposed but are seldom used, if at all.

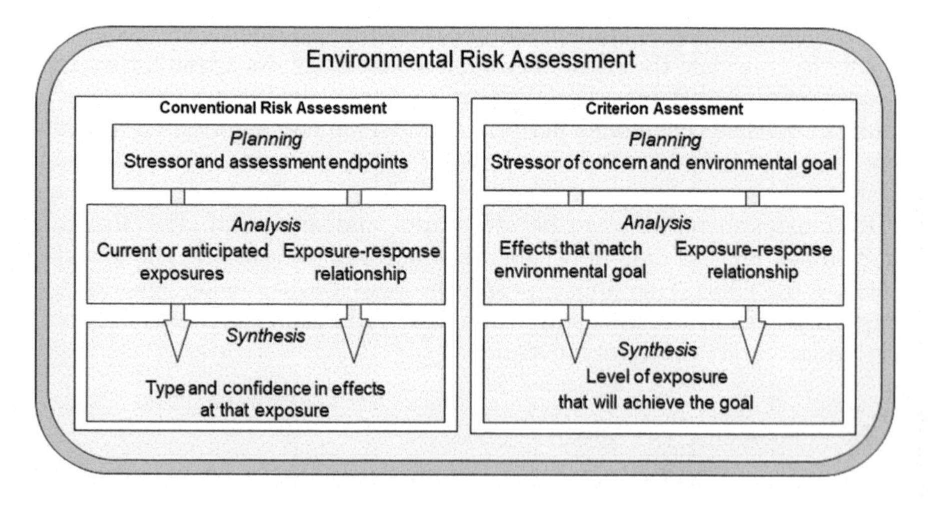

FIGURE 1.2
Frameworks for risk assessment and criterion/benchmark assessment illustrating their similarities and differences (Suter and Cormier 2008).

1.3.4 Comparative Risk Assessment

In some cases, your charge will be to choose from a set of alternative agents or actions. In the European REACH regulations, companies desiring to continue the use of problematic substances under Annex XIV must compare them to suitable alternatives. Comparative assessment is simpler than conventional risk assessment in that you do not need to estimate the magnitude of risks for each alternative. You need only rank them or, simpler still, determine which one is best because its risk is lowest. If a standard of acceptability must be met, comparative assessment is applied to the acceptable alternatives.

In many cases, the alternatives to be compared will be equivalent. For example, which detergent builder would pose the least risk to aquatic life? Because they have the same use, treatment, receiving systems, and (probably) the same mode of toxic action, such comparative assessment cases are relatively straightforward variants on conventional risk assessment. However, some alternatives are so dissimilar as to make comparative risk assessments nearly impossible. For example, alternative electrical generation technologies have myriad, incommensurable environmental risks. How do you combine the >200,000 birds killed each year in the U.S. by windmills with their various climate benefits and then compare those risks and benefits to those of gas turbines? Even if you take climate out of the assessment by comparing only to carbon neutral sources, the comparisons are still difficult. How do you compare the windmill risks to birds and bats to hydroelectric risks to salmon and other riverine biota? In such comparative assessments, you may do best to carefully describe the individual risks and leave the apples to oranges comparison to the decision-maker or even to an economist.

1.3.5 Net Benefit Analysis

You may think that remedial actions at contaminated sites are inherently beneficial. However, remediation may have serious adverse effects from digging, dredging, capping, burning (oil spills), or chemically treating contaminated media. Ancillary actions include replacing native vegetation with disturbance-tolerant species and building roads, borrow pits, landfills, and other facilities that support the remediation. Remedial actions may also cause human injuries and deaths. These actions may cause more harm than the contaminants being remediated. This issue became prominent after the Exxon Valdez oil spill in Prince William Sound, Alaska. Cleaning the rocky shoreline with a hot pressure-wash improved the aesthetics but moved the oil in a dissolved or suspended form to surface or interstitial waters and caused new biological effects (NOAA 1990). At about the same time, I was involved in an analogous case. The floodplain of East Fork Poplar Creek in Oak Ridge, Tennessee, was contaminated by mercury, but it was nearly all in mercuric sulfide and other insoluble or sequestered forms. Although it was not taken up by plants, the mercury was remediated based on a human health assessment that assumed human occupation of the floodplain and gardening. Further, the bioavailability and toxicity of inorganic mercury was assumed to be equal to the more available and toxic mercuric chloride. As a result, the wetland hardwood forest in the most contaminated reaches was bulldozed and replaced with K31 fescue, a hearty grass that can be toxic due to its endophytic fungi.

Most contaminated sites are industrial or commercial, so ecological risks from remediation are not a major issue. These "brownfields" are usually remediated to another industrial or commercial use. However, at sites like Prince William Sound and the Poplar Creek floodplain, it is important to assure that remediation will result in a net benefit. Net environmental benefit analysis (NEBA) was developed for such cases (Efroymson, Nicolette, and Suter 2004). It is a comparative risk assessment that balances risks from leaving the contaminant in place against the risks from alternative remedial actions. The challenge is estimating effects of remedial actions versus natural degradation or dilution. While many effects such as removing the biotic community and excavating the soil are obvious, others such as effects of cleaning oiled shorelines are not. Net benefits may be judged subjectively, or a common broadly applicable unit of comparison such as habitat area, number of species, or primary production may be compared. Ultimately, expert judgment is employed because common units are often incomplete summaries of effects, and diverse effects cannot be added or subtracted.

1.3.6 Economic Analyses

Two types of economic analyses are commonly applied to environmental management actions, cost effectiveness and cost–benefit analyses. They are typically applied after an environmental assessment by a team of economists, and in my experience, the environmental assessors are seldom consulted. If possible, learn what sort of economic analysis, if any, will be applied to your assessment results and what information the economists want.

Cost effectiveness analysis is intended to identify the most economically efficient way to achieve a goal. For example, if there are multiple ways to achieve a remedial goal or to meet a water quality standard, the cost of each option is determined, and the least costly option is the most cost effective one. An alternative formulation is to determine the cost of a unit increase in environmental quality for each option. That might be used to determine how to achieve the greatest improvement in some goal such as reduced area of Gulf of Mexico dead zone.

Cost–benefit analysis compares the cost to a responsible party to the benefits to human health and the environment. Economists describe the goal of cost–benefit analysis as determining whether (1) the responsible party could compensate the affected parties for loss of life, health, or environmental quality from avoided costs, or (2) the health and environmental benefits of regulation would be sufficient to allow the public to compensate the responsible party for its costs. Of course, no such compensations take place.

Cost–benefit analysis is controversial because it requires putting a monetary value on human life and health or on the environment. Costs of remediation of a site, treatment of a waste, or foregoing a market for a chemical are knowable and can be estimated by the responsible party. However, benefits are incompletely known, and because there is no market for the environment in most cases, they are difficult to reliably estimate. The most generally applicable approach is expressed preferences. People are surveyed and asked, how much would you be willing to pay for some benefit such as a 15% reduction in asthma or a 20% increase in bald eagle abundance. Less often they are asked how much they would want to be compensated for a loss of eagles or increased chance of asthma. Once again, these are fictitious transactions. I heard an environmental ethicist say that revealed preference surveys are like being asked how much would you pay me to not dump this truck load of manure on your front lawn? On the other hand, industry economists complain that people exaggerate how much they are willing to pay.

Economic analyses can be complex and arcane. I do not attempt to make environmental assessors competent environmental economists or even good judges of environmental economic analyses. At best, environmental assessors can assure that they provide assessment endpoints such as ecosystem services that are useful to economists (Chapter 14). However, economists are not limited to such endpoints. Most environmental laws in the U.S. do not set a net economic benefit as the decision criterion. However, many laws require that decisions be reasonable, and economic analyses increasingly contribute to decision making by providing one basis for judging the reasonableness of a decision. Regulatory agencies such as the USEPA provide guidance for such analyses (NCEE 2010).

1.4 Outcome Assessments

Outcome assessments answer the question, what was the result of an environmental management action? There is a temptation to issue a permit, cap contaminated soil, or build a dam, then declare success and move on. However, it is important to know whether further actions are needed to tighten the permit, remove more soil, or change the operation of the dam. For example, five years after a Superfund remedial action, monitoring data must be used to assess whether the remedial goals were achieved. Similarly, in some circumstances, after completion of a project, an audit evaluates the accuracy of the impact assessment by comparing actual to predicted impacts. Additionally, the monitoring and assessments performed to determine whether the terms of an effluent permit are being met are a kind of outcome assessment. Outcome assessments were even written into the 1987 Montreal Protocol, a treaty restricting ozone depleting chlorofluorocarbons.

Outcome assessments are often limited to a few indicators. In particular, they are often limited to determining whether exposure to high-risk chemicals or other agents has been reduced below benchmark levels. This approach is almost inevitable for human health, because in most cases, effects are undetectable because few people are significantly exposed, and among the exposed, the frequency of health effects is generally low. Ecological effects are more often measurable but are seldom measured because reduced contaminant concentrations are thought to be sufficient evidence of successful site remediation. For example, the Watts Bar Reservoir Superfund remedial investigation below Oak Ridge, Tennessee, included millions of dollars of ecological studies. However, the ecological indicator for the five-year review was the same as one for human health, just the mercury concentrations in human-edible fish filets. The lack of ecological data for the ecological outcome assessment was disappointing to me and the other ecologists who had planned to monitor recovery of the biota. However, it made sense to the USEPA region because they judged that only the human health risks were important enough to drive any further remedial actions, and they could pay lip service to ecological risks. In contrast to Superfund, the Clean Water Act has a strong mandate for ecological protection, so studies of recovery from Oak Ridge aqueous effluents and legacy contamination were substantial. Studies for the outcome assessments included chemical and biological monitoring, whole effluent and ambient toxicity tests, and toxicity identification evaluation to address causation (Greeley Jr et al. 2011). Legal mandates matter.

Although the primary reason for outcome assessments is to determine whether additional actions are required, they can also serve assessment science. That is, they can determine the accuracy and completeness of predictions, explain errors and inaccuracies, and thereby provide bases for improvement. This requires more effort to determine not only that a prediction was wrong but also why it was wrong. For example, if mercury concentrations in fish do not decline as expected, it may be due to failure to actually reduce the source, misunderstanding of the geochemistry of mercury in the river, changes in the recovering food web, or other factors.

Outcome assessments could, but in my experience never do, consider more broadly whether the assessment was effective. To what extent did the decision-maker use the assessment to guide decision making? Do the stakeholders feel that they were adequately engaged, and did they feel that their concerns have been addressed? Was the assessment cost effective? Did the actions meet relevant legal and regulatory requirements, and if not, were inadequacies of the assessment responsible? These questions are important but would require significant additional effort.

1.5 Special Purpose Assessments

You will often see reference to types of assessments that are associated with a particular problem. For example, natural resource managers must decide how to manage populations and ecosystems in the face of climate change. They may try to resist change, restore a prior ecosystem state, prevent further disturbance, accept the climate-induced changes, or manage for a new ecosystem type with desired traits. The assessments to support these decisions have been termed transformation-vulnerability assessments (Jackson 2021). However, they are the same types of assessments as described in this chapter, just with a new issue. Other examples are invasive species assessments, sustainability assessments, and nuclear winter assessments. You will find many other definitions of types of assessments that, like these examples, are characterized by their topic.

1.6 Mega Assessments

Some environmental assessments deal with major issues at regional or global scale and do not fit a usual assessment type. The obvious current example is the assessments of the Intergovernmental Program on Climate Change (IPCC). Another example is the U.S. National Acid Precipitation Assessment Program (NAPAP) which was commissioned by Congress and informed amendment of the Clean Air Act. Other examples include assessments of stratospheric ozone depletion (the assessment preceding and following the Montreal Protocol), the Gulf of Mexico dead zone (Mississippi River/Gulf of Mexico Hypoxia Task Force), and declining Chesapeake Bay fisheries (the Chesapeake Bay Program, including the bay's multi-river watershed).

Mega assessments typically combine multiple types of assessment. They often include assessments of conditions, causes, and risks and include observational and experimental studies to provide assessment-specific information. They continue for years and involve thousands of scientists. They typically involve collaborations across agencies, institutions, and even nations, but because of differing interests, the collaborations may be contentious. In the case of NAPAP, the interpretations of the science were contentious, even among U.S. federal agencies. In particular, the Department of Energy wanted to downplay environmental risks to protect fossil fuel industries, while the USEPA was more ready to acknowledge the seriousness of risks and the Reagan white house denied the reality of a problem. Meanwhile, Canada insisted that something be done about transboundary pollution and complained about the assessment process.

1.7 Summary

You should be aware of the type of assessment that you are charged with and its relationship to any prior or future assessments. In general, your assessment may determine environmental conditions, causes of impairments, effects of management actions, and outcomes of actions to address causes of new potential hazards. Each of these assessment types may stand alone or may be part of a series of assessments. Some assessments are distinctive in that they deal with distinct inferences or concern novel environmental problems.

2

The Cast of Characters

I want to be in the room where it happens.

—*Lin-Manuel Miranda,* Hamilton

Environmental assessments are performed by assessors and this book is about what you do or will do as an assessor. Few assessors have been trained in environmental assessment. I was trained in ecology and environmental toxicology and fell into assessment when promised funds for microcosm studies vanished. This chapter explains your potential roles and who you will interact with, so that you will not be as clueless as I was.

2.1 Assessors

As an environmental assessor, you bring science to bear on environmental problems by planting one foot in the fact-based world of science and the other in the values-based world of policy and decision making. Sometimes this position at the policy interface is apparent. Once, the USEPA administrator dropped in on a meeting of the Bristol Bay Watershed Assessment team. She said that the Pebble Mine permit was one of her most important decisions and ended with "Don't make me look bad." (She also took our last bagel on the way out.) Of the scientists in the Agency, it is the assessors who might cause the administrator to make a poorly informed decision that could lead to embarrassment. The same assessment was cited by Joe Biden in the 2020 U.S. presidential campaign. "It is no place for a mine," the former vice president said in a statement to news media about whether he would permit the Pebble Mine. "The Obama-Biden Administration reached that conclusion when we ran a rigorous, science-based process in 2014, and it is still true today" (Associated Press 2020). That "rigorous, science-based process" was the Bristol Bay watershed assessment for which I was scientific lead (USEPA 2014a).

The attention drawn by an assessment is not always good. For example, a Chevron Corporation vice president contacted a Department of Energy Assistant Director for Fossil Energy, who in turn contacted the Assistant Director for Life Sciences at Oak Ridge National Laboratory, who then called

DOI: 10.1201/9781003156307-3

me at home. Chevron suggested that I was unethical to suggest that oil development could harm the San Joaquin kit foxes on the Elk Hills Naval Petroleum Reserve. An investigation cleared me, and I kept my job.

Even when you are working on a lower-profile assessment, you are informing a decision that will save a stream from a toxic effluent, limit hazardous uses of a chemical, save an industry from unjustified regulation, cause a project to be redesigned to limit environmental impacts, or otherwise making a difference. To paraphrase the musical *Hamilton*, being an assessor puts you in the room where it happens.

Assessors may work for government agencies that are charged with environmental protection through regulation and management. They may also work for responsible parties including manufacturers, waste generators, and non-regulatory government agencies. In the U.S., the Department of Defense and the Department of Energy are each responsible for more contaminated land than any corporation. Assessors may be employees of the agencies and companies or of contracting firms that provide assessment expertise.

Your role is to carry out an assessment process as described in the chapters that follow. You formulate the problem; obtain, judge, analyze, and synthesize information; and produce conclusions concerning conditions, causes, or effects and confidence in those results. Ideally, you will be a generalist who can contribute to the entire assessment process but will also have a specialty such as data analysis, aquatic toxicology, statistical modeling, or environmental chemistry. You should keep up with the science, including participation on a scientific society such as the Society for Risk Analysis, Society of Environmental Toxicology and Chemistry, or Society of Toxicology.

Your role depends in part on the magnitude of the assessment. At one extreme, working alone, you may review submitted data and renew an emission permit. At the other extreme, mega assessments (Chapter 1) may have hundreds of assessors and may be very hierarchical with leaders that include senior scientists and professional managers. Unless you are one of the leaders, participating in one of these assessments can leave you feeling like a small frog in a big pond. The planning, synthesis, communication, etc. occur above your level. My part in the National Acid Precipitation Assessment Program (NAPAP) in the 1980s was to review and synthesize the literature on effects of acidification on fish. It was obviously relevant, but I did not know how it contributed to influencing Congress, if at all. In such cases, rather than communicating with the decision-maker, you communicate with the assessors at the next higher level of synthesis (in my case, the volume III, report 13, section 2 lead author). In mega assessments, assessors typically trade off making a big contribution for participating in something big. If it is big enough, you may even get a piece of a Nobel Prize, like the contributors to the IPCC climate assessments.

You may participate in the R&D component of environmental assessment in two ways. First, you may direct data generating activities to ensure that the results are useful to you. You may even participate in field or laboratory activities. Site visits are particularly important. Second, you may develop assessment methods. Although the best methods are generally accepted methods because their results are by definition generally accepted, sometimes they are not applicable. In such cases, you may innovate or search for relevant nonstandard methods. If you decide to use your innovative method, you should be cautious and test the waters to get some assurance that the results will be accepted.

Development of assessment methods may also include standardizing methods as best practices. Government agencies that practice environmental assessment have standard methods. In the USEPA, Agency-wide guidelines and less formal white papers are developed by the Risk Assessment Forum (RAF) which consists of senior assessors. Standard methods are also developed by international organizations, most notably the Organization for Economic Cooperation and Development (OECD). If you find that you have the opportunity to participate in such an organization like the RAF or OECD, go for it, even though it means more work. I found it to be a rewarding experience.

Finally, the products or actions that you assess may be legally challenged. In that case, you may be deposed and testify in court as a technical expert or witness (Chapter 30).

2.2 Assessment Instigators

If you are an assessor, you do not have the opportunity to, like a researcher, come up with a bright idea and submit it to a funding source. Rather, someone did or wants to do something with environmental implications that prompts an assessment. They may be a manufacturer who wants to assure that their product is not going to damage the environment or who must obtain approval for the product from a government agency. They may be an operator of a facility that emits contaminants to the air, water, or land. They may be a responsible party who has contaminated the environment and must remediate the contaminated site. You may be employed by the instigator or his contractor or may be employed by an agency that has regulatory authority over the instigator's actions. You may also be an assessor for a resource management agency such as the U.S. Forest Service or National Marine Fisheries Service that needs scientific analyses of their management plans. All those organizations are potential assessment instigators.

2.3 Data Generators

Data are raw materials for analyses in environmental assessments. As an assessor, you may have three relationships to data generators.

If you are dependent on information from the literature, you should decide which data from which publications meet the needs of the assessment based on its relevance and reliability. Systematic review processes are used to achieve that goal (Chapter 21). Examples of literature-dependent assessments include the derivation of water quality criteria or assessments of existing industrial chemicals. If you need clarification of a paper or if you need the original data, you may contact the authors.

Information may be generated by scientists in your organization, possibly including you. In such cases, you are likely to get the data that you want, but you should still relate it to information from other sources. It is important to not be biased in favor of in-house data. Examples of in-house data include site-specific data for contaminated site assessments and data for assessments by industry of a potential new product.

Information may be generated by scientists outside your organization for your use. Examples include data generated by manufacturers for pesticide registration or data to support renewal of an emission permit by a regulatory agency.

2.4 Decision-Makers

Your relationship with decision-makers may be fraught. The ideal scenario is that you will receive your charge from the decision-maker, discuss the specifics concerning the scope and goals, and then go off and perform the assessment. When you are through, you present the results of the assessment and answer questions, and then the decision-maker will go off and do his thing. The decision-maker may want clarification or more information. In the worst case, he will want the results changed (Chapter 20). I never had to face that.

The role of assessors in decision making is complex. We assessors should be objective scientists who do not express opinions concerning the decision unless asked (Chapter 9). However, the science must be relevant to the decision. The decision-maker accepts the scientific findings but takes into consideration legal, political, and socio-economic factors that are not strictly scientific. Assessments may seem to be decision-forcing, as in a risk assessment that finds high risks of severe effects or a cost–benefit analysis that finds high costs per unit risk reduction. However, you still should not advocate for the obvious conclusion.

To ensure that assessments are useful, assessors should keep in mind the constraints on the decision-making process.

- Legal requirements must be satisfied. So, know the relevant requirements and legal precedents.
- Multiple goals must be accommodated. So, consider how to integrate or at least compare multiple human health, ecological, and economic goals.

- Decisions must be scientifically defensible. So, use state-of-practice science and avoid errors.
- Decisions must consider available technologies. So, consider what is feasible and how remedial technologies may be effective or incidentally cause environmental damage.
- Stakeholders should be considered. So, address environmental and health issues raised by stakeholders.
- Protection should be adequate. So, be aware of how adequate protection has been defined in practice.
- Politics may intervene. There is not much you can do about this but be aware.

2.5 Stakeholders

Stakeholders (interested parties) are people or organizations with a stake in the environmental management decision. They may include a landowner adjacent to a Superfund site, the party responsible for that site, environmental and industry advocacy groups, tribes, and local governments. Conventionally, in U.S. government agencies, stakeholders may weigh in at the beginning of an assessment process to express their preferences and to contribute information. When the draft assessment is complete, the assessment is released for public comment and stakeholders, and the public may be briefed. They may express their opinions during a briefing or contribute their written comments in a public comment period. Stakeholders may take an agency to court based on alleged inadequacies or biases in an assessment. Stakeholder engagement may be contentious and large diverse stakeholder events may require professional facilitators. In the extreme, they require police or other protective services.

In general, the most active stakeholders are regulated industries. I have attended public briefings where the only participants were industry representatives. The USEPA has tried to engage a broader range of stakeholders, which resulted in accusations of subsidizing and even generating anti-industry opinion. On the other hand, the USEPA has engaged in collaborative assessments with industry groups with the intention of ensuring that the best science is applied. That practice results in accusations of collusion with the regulated party. Stakeholder engagement may shade into lobbying, with industry managers and attorneys bypassing the scientific engagement process and meeting with the politically appointed top management of the Agency.

Contrary to the expectation of the USEPA (1998) guidelines, I have learned that stakeholder engagement is seldom an opportunity for scientific dialogue.

Typically, in pre-assessment interactions, you or your management brief the stakeholders on the nature, scope and timing of the assessment and its relation to the decision. Then you receive their comments. Post-assessment, you summarize the results for the stakeholders and any implications and again accept their comments (Chapter 9).

2.6 Summary

As an assessor, you are not isolated. You are part of a community that initiates assessments, generates data, makes decisions, and works to influence the decision. In the course of a career, you are likely to play more than one of these parts.

3

Frameworks for Environmental Assessment

A framework of types of assessment problems and the relationships among the assessments is presented in Chapter 1. It shows how problems and assessments are linked and lets you know what you are assessing and why. The types of environmental assessments each have their own assessment process. If the process is mapped out, you and your colleagues know what you are doing, and reviewers, stakeholders, and decision-makers will have an idea of how you derived the assessment results. These process diagrams are commonly known as assessment frameworks. The basic framework consists of three steps plus input and output (Figure 3.1).

3.1 Process Framework for Predictive Environmental Assessments

The best-known environmental assessment framework is the USEPA's (1992) framework for ecological risk assessment (Figure 3.2), which was adapted for a cumulative assessment framework (USEPA 2003b), a human health risk assessment framework (USEPA 2014c), and an integrated environmental risk assessment framework (WHO 2001).

The impetus is the reason for the assessment, which comes from outside the assessment process. For a predictive assessment, it may be a product, project, plan, or other proposal. Examples include a highway project that requires an impact assessment, an application for a pesticide permit, a new plasticizer that may or may not be manufactured, or a gaseous emission that requires a permit. More rarely, the impetus may be objectives based. For example, assess the environmental effects of stopping the spread of an invasive forest pest or of providing rural electrification. Assessments of objectives may determine if any means of achieving the objective is environmentally acceptable or which of the alternatives is least damaging.

Any assessment begins with some sort of planning process, but until the USEPA's framework for ecological risk assessment (Figure 3.2), there was no formal process that encompassed the complexities of planning an environmental assessment. That ecological risk assessment framework divides the planning process into a planning step that involves assessors with a

DOI: 10.1201/9781003156307-4

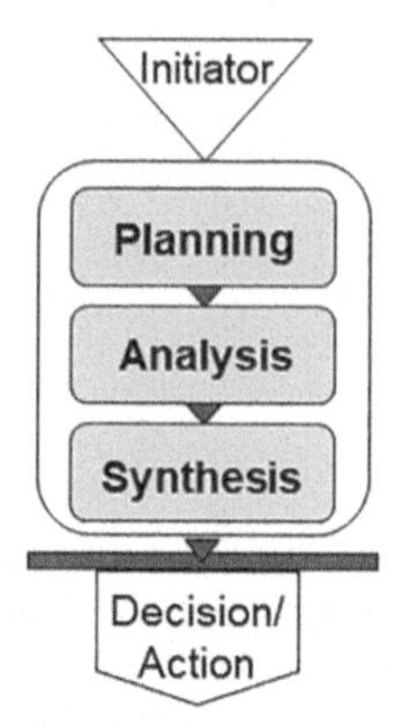

FIGURE 3.1
A basic framework for all types of environmental assessment (Cormier and Suter 2008; USEPA 2010).

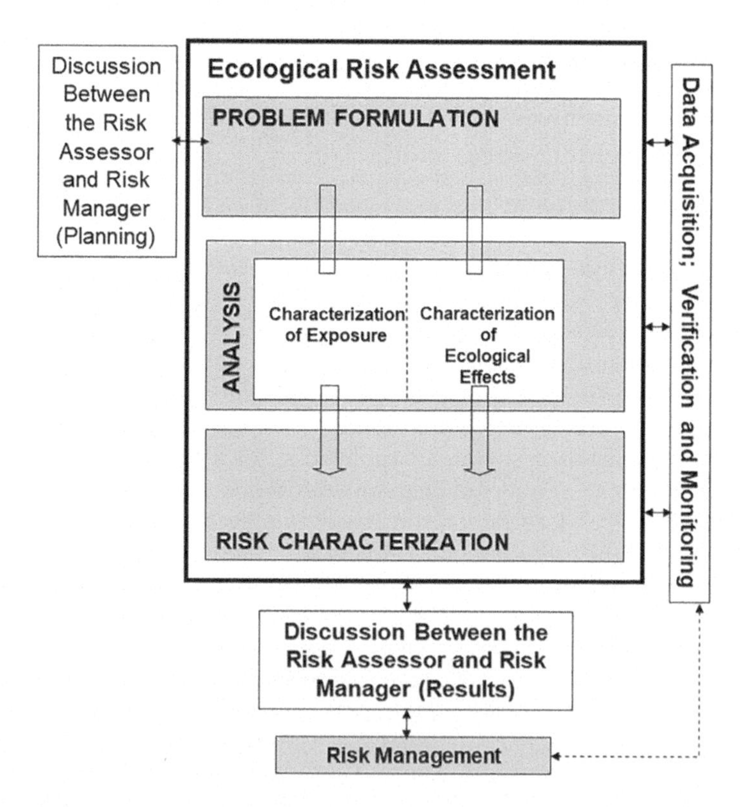

FIGURE 3.2
The USEPA (1992) framework for ecological risk assessment.

decision-maker and then a more technical problem formulation performed by assessors (Chapter 4).

The analysis phase (Chapter 5) is just what the name implies, the phase in which data and qualitative information are analyzed. In conventional risk assessments, the results of analysis are an exposure–response relationship and an estimate of exposure (Chapters 6, 7).

The synthesis phase brings together the results of analysis to determine the results of the assessment (Chapter 8). In risk assessment, this is called risk characterization, and it estimates effects by solving the exposure–response relationship for the estimated exposure. In benchmark assessments, the relationship is solved for the accepted effect level to derive a benchmark value. In causal assessments, the evidence is weighed to determine the best-supported cause. In addition to providing the results, syntheses explain the results in a narrative. An important component of the synthesis is a determination of confidence in the result and unresolved uncertainties (Chapter 17).

Following completion of the assessment, it is reviewed within your organization and sometimes externally and revised in response to the comments. Finally, depending on the institutional processes, assessors may brief managers and decision-makers, inform stakeholders and the public, or participate in the decision-making process.

Process frameworks range from the very general (Figure 3.1) to very specific for individual programs. Diagrams for specific assessment processes in specific programs may resemble a programmer's flow charts with do statements and decision gates.

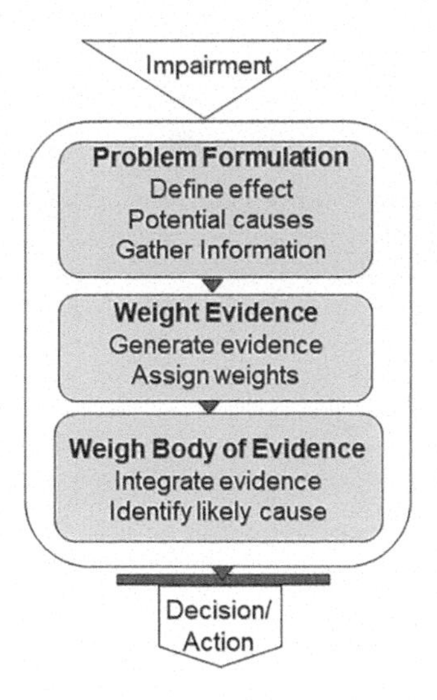

FIGURE 3.3
A general framework for causal assessment by weight of evidence.

You will probably never need to develop a process framework. You will use the framework adopted by your organization for the type of assessment at hand. However, it is important to understand the standard framework so that you can apply it to your assessment tasks. In some cases, changes in the assessment task or in organizational policies may require modification or replacement of the framework.

3.2 Process Frameworks for Other Types of Assessment

Any of the types of assessment (Chapter 1) can be framed in terms of the generic assessment framework (Figure 3.1). For example, Figure 3.3 provides a framework for causal assessment by weight of evidence. It is a simple merging of the general weight-of-evidence framework (Figure 25.1) with the generic assessment framework.

4

Planning and Problem Formulation

Any deficiencies in problem formulation will compromise all subsequent work.
—Suzanne Marcy (USEPA 1998)

To perform a useful assessment, you must determine what you are going to do and how you are going to do it. That is, you must plan the assessment. However, because environmental assessments are inherently complex, the planning process may be complex and involve numerous contributors and a lot of work. The planning effort depends on how routine the assessment is. Some regulatory assessments have established protocols. Other assessments can be modeled on precedents. For example, a new pyrethroid insecticide may be assessed like the last one. However, an assessment of a large, contaminated site, a novel agent such as a genetically altered organism, or a proposed strip mine may require much preliminary characterization of the sites, agents, and potential hazards as well as consideration of the most appropriate assessment methods.

Every type of assessment requires a different planning process. Planning for causal assessments involves identifying the adverse effect and the candidate causes. Planning for impact assessment identifies the project and the legally required considerations. Planning for risk assessments is more complex and common than other types of environmental assessments, so it is addressed at length.

The USEPA ERA framework (Figure 3.2) divides the planning process into two steps: planning and problem formulation. This distinction derives from the concern that political appointees and other decision-makers should have minimal opportunity to bias the risk assessment toward a particular conclusion. Therefore, planning is devoted to nontechnical issues and it involves managers, policymakers, and in some cases, stakeholders along with assessors. Problem formulation is devoted to technical planning and is performed by the assessment team. The distinction is not always clear, and assessors may need to consult a decision-maker and even some stakeholders at multiple points in the assessment. However, the goal of minimizing political or management bias in regulatory assessments is maintained. The issue of minimizing management bias can also arise in assessments performed for businesses and interest groups.

DOI: 10.1201/9781003156307-5

4.1 Planning

The purpose of the planning phase is to determine your charge from the decision-maker. A decision-maker such as a USEPA regional administrator or a corporate official is the ultimate audience for the assessment. However, the decision-maker is subject to whatever impetus creates the need for an assessment. For governments, these include laws and regulations, proposed projects, and applications for permits. For businesses, impetus includes product development, market demands, regulatory demands, and commitments to sustainability. Do not count on meeting the person with decision authority, because he or she will often delegate assessment planning to a lower level. Questions to be resolved by planning include:

- What are the decisions to be made and the management options?
- What are the management goals?
- What are the legal, regulatory, or policy constraints on the assessment and decision?
- Are any assessments desired beyond environmental risk such as net benefit, cost/benefit, or outcome assessments?
- What are the deadlines and resource constraints?
- Is this an internal document, or are there other audiences such as regulators, investors, or the public?

Understanding the decision to be made should include not only the laws, regulations, and policies but also the drivers of the decision. In the case of the Pebble Mine in the Bristol Bay watershed assessment, the major public issue was protecting salmon. However, we understood from the beginning that the decision driver was likely to be Section 404(b) of the Clean Water Act which regulates elimination and alteration of wetlands, streams, and other waters. The mine and infrastructure construction would inevitably result in the destruction of many square kilometers of wetlands and kilometers of streams, and mitigation would not be possible in that pristine area. Therefore, delimiting wetlands and streams in the footprint of the mine, road, pipeline corridor, and port was essential. The rest of the assessment, including risks to salmon, could have been sufficient as decision drivers, but they also served to demonstrate that it would not be in the public interest to grant a Section 404 exemption.

In the simplest cases, the results of planning are for internal use and may be in the form of meeting notes that the participants may sign to indicate concurrence. However, some contexts, such as IRIS chemical assessments, planning and problem formulation results are formal public documents (USEPA 2019c).

The planning process should focus on identifying the information needs to inform a management decision. This is based on the feeling expressed in the silver book that human health risk assessments had been focused on completing the assessment process for its own sake (NAS 2009). This focus on purpose was also expressed in the ecological risk guidelines, but less emphatically. Planning for human health assessments emphasizes some human-specific issues including environmental justice, protection of children, and sustainability.

4.2 Problem Formulation

Your first task of problem formulation is to formalize the purpose of the assessment so that you know what to analyze. Given the purpose, you proceed with the other tasks: reviewing available information, describing the environment, developing a conceptual model, determining the assessment endpoints, and deriving an analysis plan for the assessment.

4.2.1 Hypothesis, Question, or Hazard

What is the subject of your assessment? Depending on the context and nature of the assessment, it may be a hypothesis to be evaluated, a question to be answered, or a hazard to be characterized. For example, a hypothesis may be that "chlorine in the effluent is the cause of low invertebrate abundance in the receiving stream." Questions might be "what is the cause of low invertebrate abundance" or "what chlorine concentrations will not cause biological impairment in a stream?" A hazard might be "is the permitted level of chlorine in treated sewage effluents a hazard to aquatic invertebrates?" The differences in these expressions of the subject of an assessment are largely a matter of appropriateness to the type of assessment.

In some cases, your subject is sufficiently specified by the charge to the assessors from the planning process. For example, determine whether chemical *X* is a carcinogen, determine the potential impacts of a proposed mine, or determine the cause of a fish kill. However, it is common to be charged to assess an agent or project and you must determine what ecological or human health hazards it poses. In such cases, the conceptual model and common knowledge may be sufficient to identify the hazards. However, if the hazards are not evident, a review of the literature or of data generated for the assessment is required (Section 4.4). If that is not sufficient, the hazard should be identified during the analysis of effects (Chapter 7). In some cases, the best that can be done during the problem formulation is to identify a very general hazard such as "aquatic toxic effects" or "noncancer human health effects."

There may be more than one subject of an assessment. For example, you might be charged to answer a qualitative question (what is causing the biological impairment?) and a follow-up quantitative question (what should be the total maximum daily load of the causal agent? (https://www.epa.gov/tmdl)). Commonly, an assessment of a project or contaminated site will assess multiple hazards due to multiple activities or contaminants or in different locations.

4.2.2 Information Review

To formulate the assessment problem, you must determine what is known about the agents or actions to be assessed, potential receptors, and receiving environment. Regulatory agencies often receive information from the manufacturer or other responsible party. Someone must assume the burden of reviewing the literature, interviewing plant managers and waste disposal operators, obtaining data from investigators, and in other ways, tracking down relevant information. The information must then be compiled, organized, and summarized.

In recent decades, literature reviews have been formalized into systematic reviews (Chapter 21). The traditional literature review practice is to simply search on some key words, scan reference lists, and contact experts until you feel that you have done enough. That approach is not replicable, is likely to miss important information, and may produce a biased set of information. Systematic reviews involve a detailed plan, search strategy, screening and acceptance criteria, and documentation of the process and results. They are transparent and defensible but can be onerous. However, processes are being developed that use problem formulation to inform the systematic review and the results of the systematic review to refine the problem formulation (Roth, Sandström, and Wilks 2020).

Often, the most useful information comes from assessment-specific studies. In some cases, such as pesticide registration, the applicant supplies a substantial information package to the regulatory assessors. Depending on the case, you may have the opportunity during problem formulation to request measurements, observations, or tests that are needed for an adequate assessment. These should be included in the analysis plan. Assessments of contaminated sites are based on measurements of the contaminants in water, soil, sediment, air, or biota and in the waste itself. They may also include biological surveys to determine what effects are associated with the contamination. Finally, contaminated media may be sampled and tested for toxic effects.

4.2.3 Environmental Description

You should describe the environment which is the setting for the assessment. The choice of environment depends on the assessment.

- A real place of concern such as the area that has been contaminated by a pipeline spill.

- Real representative places such as typical streams that would be crossed by a proposed pipeline.
- A worst real case such as use of pyrethroids on cotton in Yazoo County, Mississippi (Hendley et al. 2001).
- A hypothetical place which is representative of a region or of the most sensitive places or has default environmental attributes.

The environmental description in the problem formulation may be introductory. Because they are used in modeling or other inferences, the details of physical, chemical, and biological conditions are often determined in the analysis phase.

4.2.4 Conceptual Model

Conceptual models are diagrams or other graphics that represent the components and potential causal structure of the system to be assessed (Chapter 23). For assessments of environmental contaminants, they represent the sources, releases, transport, transformation, exposures, proximate effects, and direct and indirect endpoint effects (Figure 4.1). Indirect effects on a species are those such as loss of food or habitat that result from the direct effects of the contaminant on another species. Figure 4.1 shows a standard format for Superfund remedial investigations which includes human health and

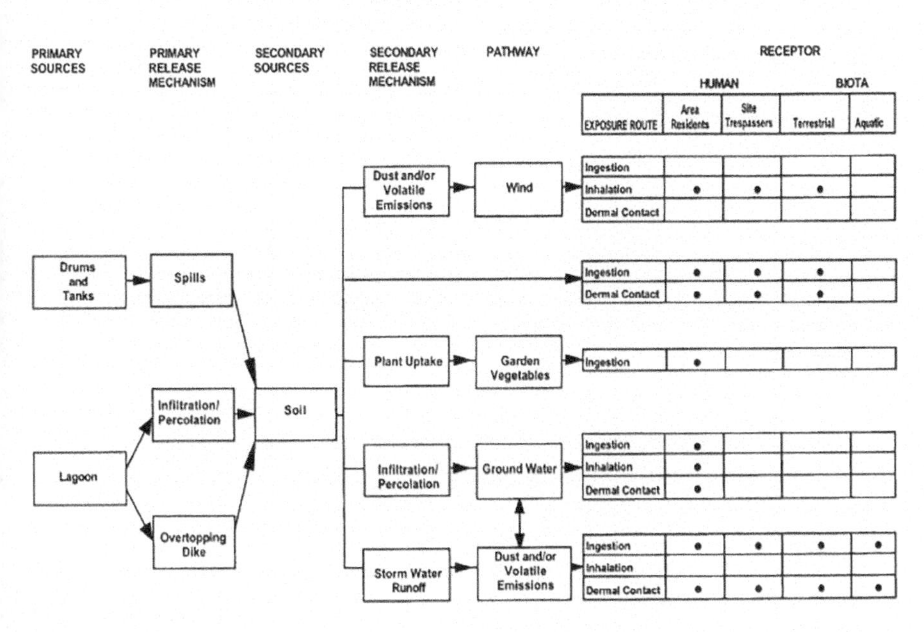

FIGURE 4.1
A standard form of conceptual models for assessment of risks from waste sources to humans and ecological receptors (Office of Emergency and Remedial Response 1988).

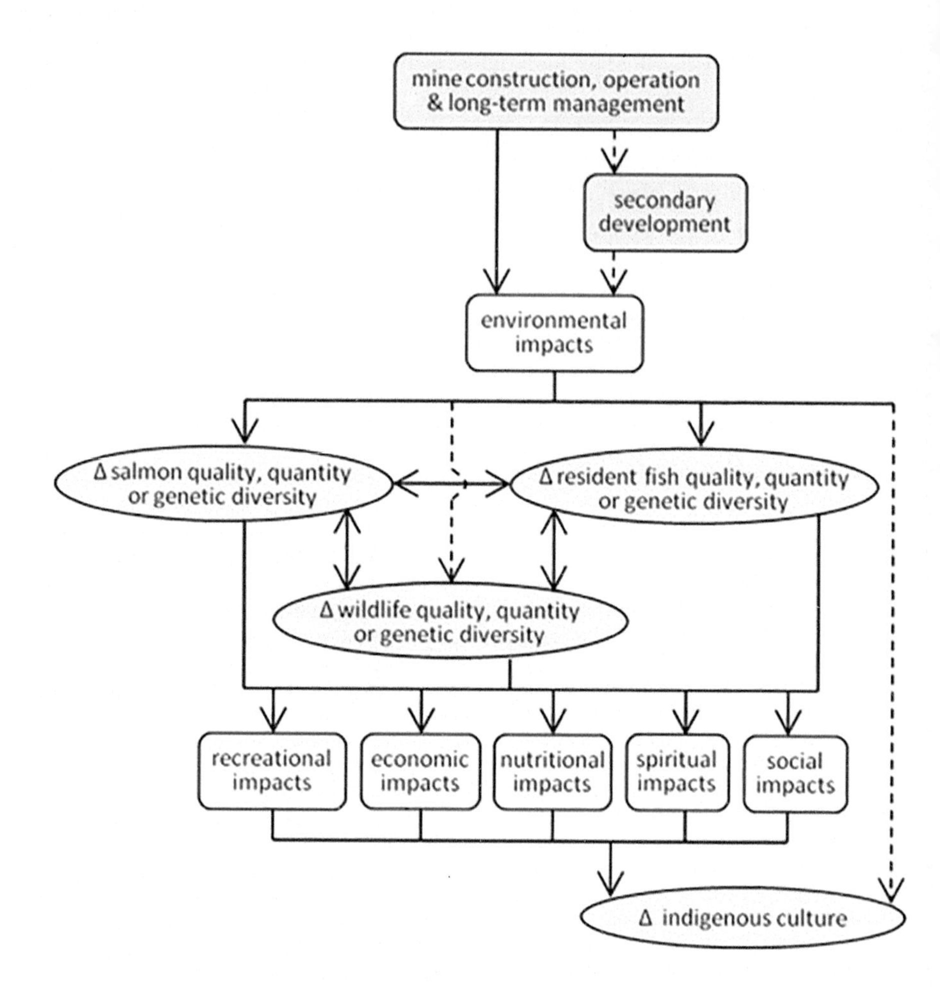

FIGURE 4.2
An aggregate conceptual model of the potential effects of a proposed open pit copper mine in the Bristol Bay Watershed, Alaska (USEPA 2014a). Because the complete conceptual model requires four fold-out pages, this summary model was used to show the overall structure of the causal relationships.

ecological risks. Conceptual models for other agents, such as dams, biocontrol agents, or prescribed burns, would have analogous structures showing how a human action is causally linked to effects of concern (Figure 4.2). This conceptual model for the pebble mine proposed for the Bristol Bay watershed is an aggregate summary of the four large and complex conceptual models needed to portray the effects of mining in a roadless area on natural and human communities (USEPA 2014a).

Developing a conceptual model helps an assessment team to understand the system and determine what must be included in the analysis. Conceptual models also serve to guide the development of quantitative models and the characterization of analytical results. Finally, they aid communication with stakeholders and decision-makers.

4.2.5 Assessment Endpoints

When planning a risk assessment, you must know what result you want in the end; what effects on what valued environmental attributes must you characterize (Chapter 14)? These assessment endpoints must be consistent with the hypothesis, question, or hazard, but more specific. One of the innovations of ecological risk assessment was to clearly distinguish the assessment endpoints from endpoints of toxicity tests and other studies (Suter 1989). The latter are termed measurement endpoints. For example, the assessment endpoint for an effluent may be the species richness of fish, and the measurement endpoint may be an LC_{50} for fathead minnows.

The assessment endpoint is defined by an entity and an attribute of the entity. In the example, the entity is the fish community in the receiving stream and the attribute is the species richness. The USEPA's three criteria for ecological endpoints are relevance to management goals, susceptibility to known or potential agents, and ecological relevance.

The selection of assessment endpoints is aided by the USEPA's (2003c, 2016c) generic ecological assessment endpoints. They include both conventional endpoints and ecosystem services. The primary advantage of these generic endpoints is that they are assured of being relevant for assessment of environmental contaminants in the U.S. based on law, policy, and precedent. However, these broadly defined endpoints must be refined for individual assessments. In particular, case-specific exposure and susceptibility make some specific endpoints more appropriate than others. Assessments of herbicides include attributes of algae and other plants. Assessments of insecticides include attributes of valued nontarget insects such as bees, butterflies, and predatory beetles. For example, an assessment of imidacloprid focused on attributes of honeybees which were thought to be declining, in part because of neonicotinoid pesticides (OPP 2020). Assessments of antibiotics in the environment consider effects on both important microbial processes and the acquisition of resistance by pathogens. In addition to susceptibility to the agent, the routes of exposure and the ecological and management relevance of the potential endpoint should be considered.

4.3 Dimensions of the Assessment

Your environmental assessment will estimate or describe something. It may be the area of habitat lost, the permissible concentration of a chemical in water, the permissible release rate from a source, the proportion of species lost below an outfall, the nontarget species susceptible to a biocontrol agent, or something else. Although it is not a consistent practice, you can clarify the results of the problem formulation by specifically defining the full set of dimensions of those assessment results. Most or all of these will be attributes of the assessment exposure or assessment endpoint. The USEPA's cumulative risk assessment and HHRA guidelines associate dimensions with steps in a generic conceptual model. The EFSA Scientific Committee (2016), for each ecological endpoint, defines five dimensions: ecological entity, attribute, magnitude, temporal scale, and spatial scale. The EFSA dimensions are generic and defined categorically. For example, the scale for magnitude is negligible, small, medium, and large.

Dimensions proposed for the U.S. National Ambient Water Quality Criteria (NAWQC) are more specific (Suter 2018).

1. The assessment endpoint consisting of an entity that is susceptible and worthy of protection, and an attribute of the entity that is affected (Chapter 14).
2. The threshold level of effect on the endpoint attribute is defined by a magnitude of effect or the proportion of entities affected.
3. Temporal dimensions may include the duration of the exposure and the frequency with which exceedances of a standard may occur.
4. Exposure metrics may include total concentration, dissolved concentration, free ionic concentration, temperature, or specific conductivity.
5. Limits of applicability are defined on chemical, spatial, or ecological dimensions such as salt water versus fresh water.

4.4 Assessment Approach

What general approach will you use? The options are numerous, but the following are common general approaches. Your job is to decide what approach is fit for your assessment purpose.

Comparison to benchmark values. In simple cases, you compare measured concentrations or equivalent exposure values to benchmark values. The benchmarks may be permit levels or enforceable standards, and the purpose of the assessment may be regulatory, in which case standard procedures are used. Alternatively, they may be used in a condition assessment to determine whether there is a problem or in a screening assessment to determine whether a contaminant is worth further investigation. In such cases, you must determine the appropriate comparisons and judge whether any available benchmark is appropriate to your assessment.

Predictive modeling. You may use estimated releases of a new chemical in a multimedia model to estimate environmental concentrations or doses for humans and other species. You may then estimate effects type, severity, and distribution of effects from a model of an exposure–response relationship. In complex cases such as a new mine, you may use models to predict effects on air and water quality, habitat loss, and finally effects on biotic communities.

Environmental epidemiology. You may apply epidemiological assessment methods when a human or nonhuman population displays an adverse effect or is known to be exposed to a hazardous agent. That is, by surveys and measurements, quantify the exposures and effects in the populations of interest. The degree of association of exposure and effects and other evidence are used to infer the reality of the apparent effects and their causation.

Tiered assessment. When you can generate or demand information, you can use a system of tiered testing and assessment (Figure 4.3). The first tiers are quick and cheap and are intended to identify uses that are clearly acceptable or unacceptable. If the results are ambiguous, you can use more costly and time-consuming tests and models to increase confidence and reduce uncertainty. Eventually, the amount and realism of information generated should allow you to determine whether the estimated effects of exposure are acceptable. This approach can be very efficient, because in most cases the risk is clearly acceptable or not from a simple assessment. As a result, the most common tiering is a screening assessment and a definitive assessment. For example, most contaminated site assessments begin with a screening assessment that uses protective screening benchmarks and simple high-end exposure assumptions. For any contaminants that are not screened out, less conservative methods are used to estimate effects and associated confidence.

Weight of Evidence. Often, but particularly in causal assessments, multiple studies and multiple types of evidence are available to perform an assessment. In such cases, you can weigh the body of evidence to determine the best supported hypothesis, level of effect, or benchmark value (Ch. 24).

Comparison of Alternatives. You may be called upon to compare of alternatives, such as which fire-fighting foam poses the least risk? In the U.S., comparisons of alternatives are required for environmental impact statements and Superfund remedial investigations. Comparative assessments require more information and analyses to cover the alternatives, but the inference is conceptually easier. Rather than reaching an absolute judgment which requires estimating the type and level of effects, you can simply conclude that one alternative is better. For example, you may choose the alternative that destroys the smallest area of vegetation, without determining the magnitude of effects on biodiversity in an area lost. Similarly, for human health assessments, you may choose the least toxic chemical for a use without estimating the number of people who would be sickened by each alternative chemical.

Comparative assessments of alternatives require that you specify the dimensions of the alternatives to be considered. You might compare alternative technologies, locations, temporal or spatial extents, or intensities of an action. Alternatives may be formulated as scenarios. A conspicuous example

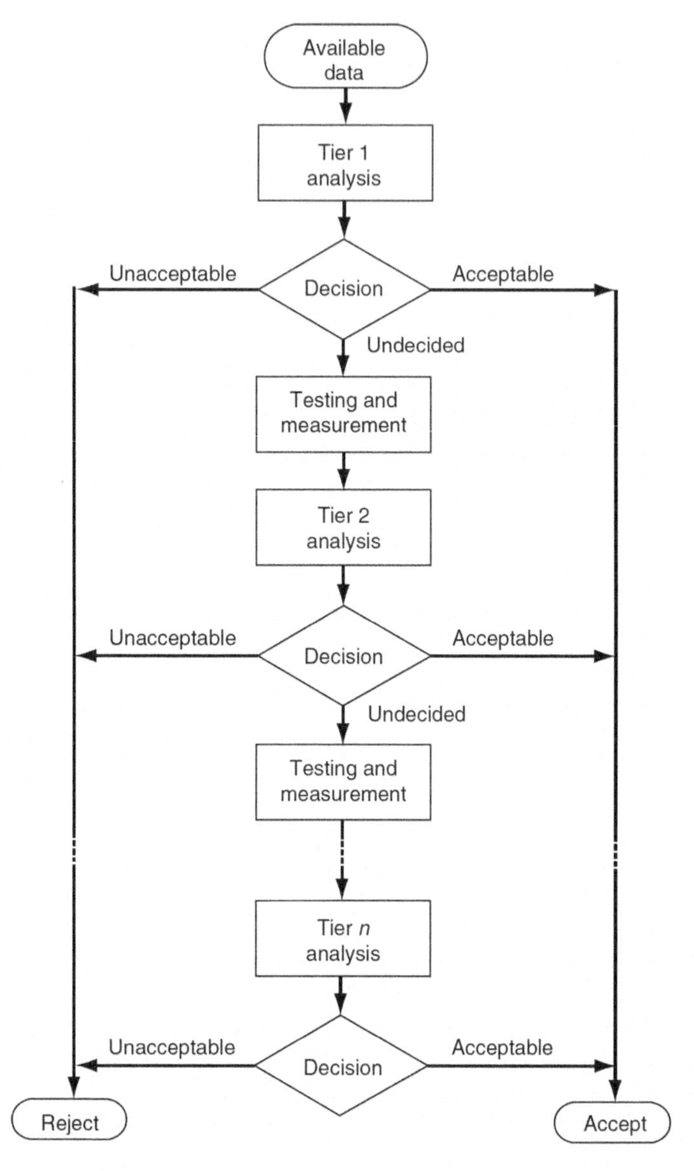

FIGURE 4.3
A framework for assessments based on tiers of testing, measurement, and analysis. Early tiers are quick and easy, and succeeding tiers are more complex and costly and provide more information. (From Suter 2007.)

is the IPCC's (2000) scenarios for alternative climate futures. IPCC estimated future carbon emissions based on assumptions about regionalization versus globalization and economic versus environmental focus. These emission scenarios are used to compare effects on the environment, human health, and socio-economic conditions with respect to the alternative policies.

4.5 Analysis Plan

The analysis plan determines how the analysis and synthesis phases of your assessment will be conducted. Some assessment processes are standardized, so you simply consult the established protocol. Otherwise, the assessment team must agree on how they will proceed to perform the analyses and characterize the risk, impacts, or other assessment results. The analysis plan is based on the results of the prior steps in planning and problem formulation. Depending on the context, your analysis plan may be a detailed procedure or a general consensus approach that you flesh out during the analysis and characterization phases.

One of the most important parts of developing the analysis plan should be coordination among the various specialties. Environmental chemists and toxicologists must agree on how exposure will be expressed so that the exposure estimates, and the exposure–response models will be concordant. Ecological and human health assessors should agree on assumptions so that their results will not be contradictory. Statisticians and other assessors should agree on expressions of confidence, uncertainty, and any probabilistic results.

If data generation is an option, your analysis plan includes an information shopping list. What can and must be tested, measured in the lab or field, found in the literature, defined by default assumptions, modeled empirically, or mechanistically simulated? What data quality is required? Will models be verified? What other considerations will apply to your assessment?

General issues for the analysis include, will you use a tiered approach or a single analysis, will you weigh all evidence or a follow single line of inference, how will you deal with confidence and uncertainty, and do you intend to generate new data? If field data are used in the analyses, it is important to identify standards for reference sites and to consider how to deal with potential confounding variables.

4.6 Human Health

The concept of problem formulation was developed for ecological risk assessment (USEPA 1992) and eventually included in human health risk assessment

(USEPA 2014c). Because HH assessment is generally less complex, the HH problem formulations are simpler. They deal with the same species every time, the same level of organization, and a few modes of exposure. Ideally, ecological and HH problem formulations are performed jointly so that they are consistent (Chapter 13).

4.7 Review of the Plan

Because the problem formulation determines how the assessment will be performed and what the ultimate results will be, it may be reviewed by the decision-maker. In my experience, decision-makers do not expect to review the problem formulation, so it is up to the assessors to raise the issue or not. Asking, "is this what you want?" may have advantages and disadvantages. On the one hand, a decision-maker's review may make it more likely that your assessment will influence the decision. On the other hand, a decision-maker's review may result in interference with the science, based on political or business motives. Depending on the organization and decision context, your choices may be limited. The practice of peer review and even public review of the problem formulation has appeared in a few contexts such as USEPA IRIS assessments. This can add months to the timeline and introduce comments based on ulterior motives. External review of a problem formulation is a management decision.

4.8 Summary of Problem Formulation Results

A problem formulation defines what you know, what you need to know, and how you will come to know it. You must answer five major questions:

- What is the impetus for the assessment? This is answered by consulting with the decision-maker to determine the legal mandate and goal.
- What hypothesis, question, or hazard is the subject of the assessment?
- What is the system in which the agent or project interacts with the environment? The primary tool for answering this question is the conceptual model, but it should also describe the particular place, if relevant, and the types of ecosystems involved.

- What should be protected? This is answered by defining one or more assessment endpoints.
- What may be managed? This is answered by defining sources, activities, or contaminated media causing the assessment exposures.
- How will the assessment be performed? This is answered by identifying the general approach and developing the analysis plan.

5

Analysis

Get the habit of analysis—analysis will in time enable synthesis to become your habit of mind.

—*Frank Lloyd Wright (1867–1959)*

In the analysis phase of an assessment, you characterize and quantify the information required to determine the risk, impact, condition, cause, or other assessment result. In frameworks for ecological and human health risk assessment, two types of information are specified for analysis, exposure and effects. If, however, you look at your conceptual model and think about how the world works, you realize that the distinction in the framework is only a convenience. In the real world, one thing causes the next thing, which causes the following thing, etc. Therefore, the effect at one step in a causal process may become the exposure in the next step.

Consider Figure 5.1. It is a very generic conceptual model for direct toxic effects of environmental contaminants. The source strength is the ultimate characterization of exposure—the environment is exposed to releases from the source. At the other end, the assessment endpoint is the ultimate characterization of effects. To distinguish the analysis of exposure and analysis of effects in the assessment process, you divide the cause–effect continuum at one of the dashed lines. What is at the dashed lines? It is exposure–response models derived from field or laboratory tests or observational studies (Chapter 7). These models of the relevant causal relationship are the reason we must distinguish exposures and effects. That is, you must have an exposure estimate (shown just before a vertical line) that has been derived from the analysis of exposure. Then you use that exposure to estimate the effects (shown by everything after that vertical line) by solving the exposure–response relationship from the analysis of effects for the exposure.

It has long been recognized that there are two distinct endpoint effects in environmental assessment (USEPA 1998). The one that is of interest to the decision-maker and is the object of the assessment is the assessment endpoint (Chapter 14). The one that is derived from measurements in a test or observational study and appears in the empirical exposure–response model is the measurement endpoint. If your measurement endpoint is not the same as the assessment endpoint, you estimate the assessment endpoint from the measurement endpoint by extrapolation (Chapter 26). For example, you may have a LC_{50} (median lethal concentration) for fathead minnows as your only

DOI: 10.1201/9781003156307-6

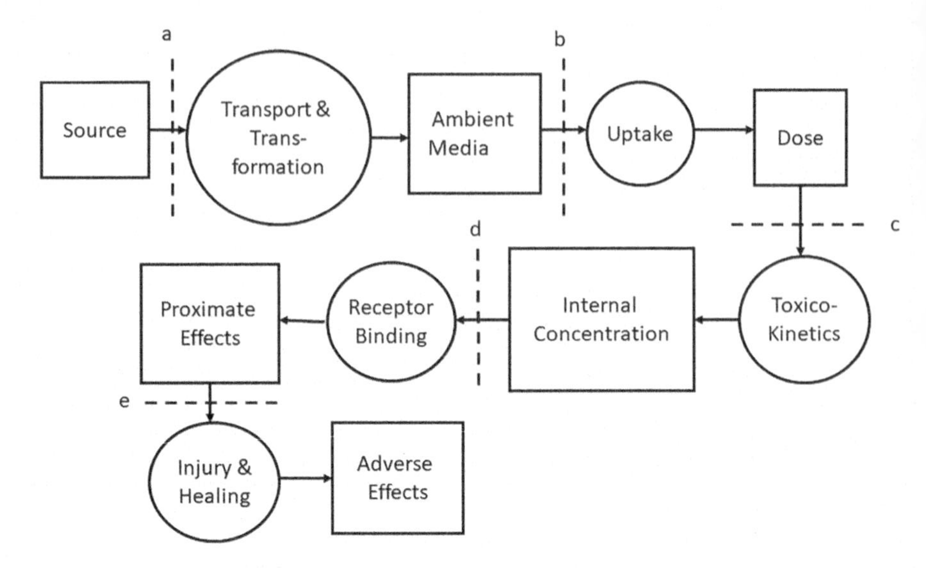

FIGURE 5.1
A generic linear conceptual model from source to endpoint effect. Each dashed vertical line is a possible point at which you might divide the results of an analysis of exposure from the analysis of effects. The assessment endpoint effect is always the adverse effect at the end. The model shows alternating states (rectangles) and processes (circles).

measurement endpoint, but your assessment endpoint is death of any species of fish in a receiving community. You may estimate that assessment endpoint by dividing the LC_{50} by a factor of 100 (Chapter 26).

It is less well recognized that there is an equivalent distinction on the exposure side between what is measured and what we want to know. The exposure that is regulated, remediated, or otherwise controlled and therefore is an object of the assessment may be termed the assessment exposure. The one that is derived from measurements and appears in the exposure–response model is the measurement exposure (Box 5.1). Ideally, the two exposures are the same, but often they are not. When the measurement and assessment exposures are not the same, or when there are no measurements of exposure, what is known and what is needed must be bridged by an exposure model. For example, when assessing a source effluent, the assessment exposure is the concentration in the effluent because the effluent is regulated. However, the effects on the aquatic biota may be determined from an exposure–response relationship solved for the concentration below the zone of initial dilution (the measurement exposure). A dilution factor (a very simple exposure model) converts the measurement exposure, which is a concentration allowing for reasonable dilution that causes an acceptable effect, into an assessment exposure which is the maximum acceptable effluent concentration.

These distinctions can be illustrated by relating exposures and effects to the dividing lines (a-d) in the causal chain in Figure 5.1.

> **BOX 5.1 EXPOSURE TERMINOLOGY**
>
> I created the measurement exposure and assessment exposure terminology for this book to clarify the types and uses of exposure. Do not expect to see it in other documents. Use it if you find it to be useful.

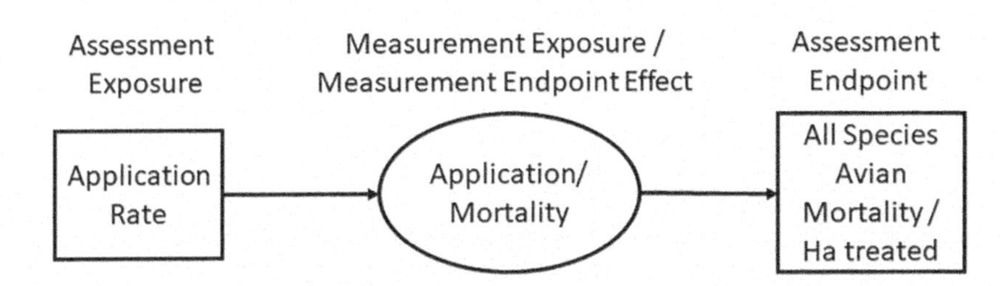

FIGURE 5.2
A linear conceptual model for cases in which the assessment exposures and measurement exposures are the source strength. The source in this case is the Application Rate, the adverse effects are All Species Avian Mortality, and everything in between is collapsed into the oval box's process model.

a. You may define assessment exposures and measurement exposures as the source strength. For example, acceptable pesticide exposure is commonly defined by the application rate (kg/ha spayed on a crop). If field studies are performed, measurement exposures are the application rates, so the measurement and assessment exposures are the same (Figure 5.2). Similarly, avian mortality may be observed in the sprayed area, so the measurement endpoint effect and the assessment endpoint effect are the same (e.g., dead birds per hectare). However, if field studies are not performed, an exposure model must estimate avian doses (the measurement exposure in laboratory toxicity tests) from application rates (still the assessment exposure) (Figure 5.3). Similarly, proportional mortality of a test species is the measurement endpoint (e.g., LD_{50}). That measure of effect and extrapolation models are then used to estimate mortality of all species to derive total avian mortality per unit area treated.
b. You may define the assessment exposure as the concentration of the substance in an ambient medium. This is the most common assessment exposure definition in ecological and in human health assessments of contaminants in air, water, and food. For example, state water quality standards are aqueous concentrations. If observed concentrations are too high, source strengths (assessment exposures) must be identified that will achieve the standard.
c. You may define exposure as the dose of the substance taken up or ingested. This is the conventional measurement exposure metric for

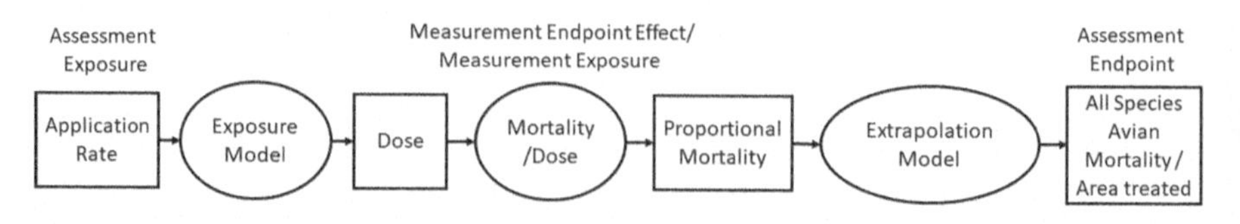

FIGURE 5.3
A linear conceptual model for cases in which the assessment exposure is the application rate and the measurement exposure is the modeled dose. The process model in the middle derives measurement endpoint effects from dose, and then an extrapolation model converts the laboratory derived measurement endpoint effect into the assessment endpoint.

human health and wildlife assessments. The exposure–response relationships are effects as a function of dose derived from laboratory tests or estimated by exposure models in epidemiological studies. However, it is not an assessment exposure, because regulatory or remedial actions are based on source strengths or ambient concentrations. Dose is a convenient metric to integrate various oral intakes and it is the exposure metric in mammalian and avian toxicity tests, but exposure modeling must be used to relate it to something that can be controlled.

d. You may define exposure as an internal concentration of the substance taken up or ingested. This is an uncommon measurement exposure, but some studies permit derivation of internal exposure–response relationships. As with oral doses, internal concentrations must be related back to a useful assessment exposure that can be controlled.

Now that you know how the causal conceptual model relates to the analysis of exposure and effects in the assessment process you can proceed with the analyses. The assessment endpoint effect and the assessment exposure should have been defined during the problem formulation. That is, you know what type of exposure may be controlled and what effects are to be prevented or remediated based on the type of assessment, relevant policies, and the decision-maker's needs. Within those bounds, you must identify the point at which you can derive the most useful and scientifically defensible exposure–response relationship which can be solved for a reliable measurement exposure. If that measurement exposure does not correspond to the assessment exposure, you need an exposure model to connect them. If the measurement endpoint effect in the exposure–response model does not correspond to the assessment endpoint, you need an extrapolation model to estimate the assessment endpoint.

6

Analysis of Exposure

Environmental assessments are concerned with the effects of some sort of exposure. Exposure is the contact or co-occurrence of a chemical or other agent with a receptor. As explained in Chapter 5, the exposure can be anything from an airplane spraying a field to a chemical binding with a receptor on a cell surface.

We assessors commonly think about exposure as the exposure metrics in exposure–response relationships derived from toxicity tests, observational studies, or models. However, as introduced in Chapter 5, the exposure that is ultimately important to the assessment is the one that will be regulated, remediated, or otherwise managed. For example, in a cadmium-contaminated stream, we express exposure as dissolved concentration to relate it to the cadmium water quality criterion. However, we regulate or manage cadmium emissions from industrial effluents, mine wastes, or other sources. By analogy to assessment endpoints and measurement endpoints, we can call the source emissions the assessment exposure and the measured concentration that is related to the criterion is the measurement exposure. In some cases, the assessment and measurement exposures are the same. However, a common measurement exposure is dose, which cannot be measured in the environment and is not regulated or remediated. Exposure models must translate between dose and something in the environment that may be managed. Assessment exposures are defined by management-relevant attributes including concentration, release rate, duration, volume, and form.

Exposure analysis may be mechanistic, relying primarily on the modeling of scenarios. Examples of such cases are new chemicals or a proposed project. Alternatively, exposure analysis may be empirical, relying on data from an existing source, contaminated medium, or physical disturbance. If a potential source is being assessed, the exposure model will begin with an estimate of the release rate from a source and simulate processes of dilution, transport, transformation, contact, uptake, and food web transfer. In contrast, exposure may require no modeling. For example, receptor organisms, such as fish in a contaminated stream, may be collected and analyzed so that exposure is expressed as whole-body concentrations for ecological assessments and filet concentrations for human health assessments. Even simpler, an exposure analysis may be simply the measurement of a concentration in water that will be compared to a water quality benchmark.

DOI: 10.1201/9781003156307-7

6.1 Exposure Scenarios

Conventionally, the exposure process is divided into source, transport and fate, and proximate exposure. The full process is defined by an exposure scenario which is your description of the ways in which the endpoint entities become exposed to the agents of concern. These include chronic scenarios and acute scenarios. The chronic scenarios usually result from normal conditions, and the acute scenarios may result from the occurrence of extreme values of the exposure processes (maximum scenarios) or from a failure of the system (accident scenarios).

6.1.1 Normal Exposure Scenarios

Exposure scenarios have two forms. First, because there may be multiple sources, pathways, and endpoint receptors, each relevant exposure pathway through the conceptual model (Chapter 23) is identified. Second, using the conceptual model as a guide, a narrative describes those pathways and conditions under which they can occur. Normal exposure scenarios are based on expected or typical conditions and parameter values. These are chronic scenarios in the sense that we expect normal conditions to occur most of the time.

Assessments often must address aggregate exposures which include the combined exposure of an endpoint entity (individual, population, or community) to a specific agent by all relevant routes from all relevant sources. Aggregate exposure scenarios should include both natural and anthropogenic sources of naturally occurring agents. They should also consider multiple routes of exposure including water, food, air, and contact with contaminated media. For example, polycyclic aromatic hydrocarbons (PAHs) occur naturally in smoke from wildfires and anthropogenically in runoff from asphalt pavement and in smoke from incinerators and diesel engines. They are released to air and water and transfer to other media including animal tissues. In such cases, you may calculate the exposure from the source being assessed and simply acknowledge the other sources, incremental exposure from a new source under evaluation, or the total exposure from all sources.

6.1.2 Maximum Exposure Scenarios

Maximum exposures represent the worst credible circumstances with respect to variability of the source, transport, or receptor. Assessments of human health risks from specific local pollution are often assessed based on a maximally exposed individual (sometimes, the qualifier "reasonable" is added to the description). For air pollution from a facility, it may be the "porch potato." That is, a person with a home at the fence line of the contaminant source who spends his days on the porch breathing the polluted air as measured by perimeter monitoring or estimated by modeling. For Superfund assessments on the Oak Ridge Reservation, one scenario was the intruder who fished and hunted deer and waterfowl and who ate the fish and game. Measured radionuclide levels in fish, deer, and waterfowl were used to calculate the intruder's exposure. A version of this scenario was found to be real when it was discovered that guards patrolling the reservation at night were poaching deer caught in the spotlights. A more extreme scenario was the homesteader who settled on the site after the government lost control, dug a cellar into soil contaminated with buried waste which was spread on the surface where a garden was planted, a milk cow grazed, and children ate soil.

Maximum exposures for ecological assessments may be analogous. Individuals of a species or community may be assessed at highly contaminated locations. For example, in the San Joaquin kit fox assessment, I assessed risks to foxes from direct exposures to wastes including spilled chromate, oil spills, drilling muds, and produced water lagoons in their territories. Maximum exposures of aquatic communities to effluents may be based on highest wastewater concentrations, undiluted effluent concentrations, or concentrations at low stream flow.

Maximum exposures may be associated with an event rather than maximum chronic uses or conditions. For example, migratory birds are killed when they land in oil sumps, mine pits, or waste ponds. In a notorious case, flocks of snow geese died in 1995 and 2016 when they landed in the Berkeley Pit, an abandoned open pit mine in Butte, Montana, and drank the pH 2 water containing high levels of copper, cadmium, and arsenic. These avian exposures are maximum but not necessarily unlikely.

Maximum exposure scenarios are handy if the estimated effects of the scenario are negligible. Then you can say, because the maximum exposure is safe, a typical case is definitely safe. However, if there are significant effects in the maximum exposure, then you need to clarify the likelihood of the maximum scenario and compare it to effects of an exposure scenario for a typical human or other organism. Development of maximum exposure scenarios requires balancing imagination, realism, and the desire to ensure protection.

6.1.3 Accident Scenarios

Accidents involve failures of human-produced systems. Examples include failures of waste treatment systems, spills of oil and other products, dam collapses, and fires at production facilities. As part of any environmental assessment, you should consider what accidents may result in environmental damage. For the Bristol Bay Watershed assessment, accidents considered were failures of the earthen tailings dams, culvert washout, and rupture of the diesel fuel and product slurry pipelines. The mining company insisted that none of these could happen at a modern mine, but an earthen tailings dam at a similar active mine (Mount Polley in British Columbia) failed catastrophically a few months after our assessment was published. Chemicals can be released by accidents. For example, a detergent builder might be spilled in a tank truck wreck or an accident at a detergent formulator's plant and would be released to surface waters if a sewage treatment plant failed. Accident assessments are often stand-alone products, but a full risk assessment requires inclusion of the effects of an accident occurrence. For the Bristol Bay assessment, we estimated how much salmon habitat could be physically damaged or toxic if it was exposed to a dam, pipeline, or culvert accident.

The best source of accident scenarios is accidents that have occurred in systems like the one that you are assessing. For the Bristol Bay example, there were histories of earthen dam, culvert, and pipeline failures. If the case is novel, you must use your knowledge and imagination. The detergent builder example suggests that the life cycle of a chemical or other product should be considered when defining accident scenarios. For designed systems like water treatment plants, engineering risk assessment techniques may be used.

Fugitive emissions are a type of accidental release that is not apparent. Conventionally, they include gases or vapors from a pressurized containment such as appliances, storage tanks, or pipelines. Similarly, slow leaks of waste waters, fuels, and other liquids may be fugitive but they are not as likely to be unobserved. Fugitive emissions are often small, but over time they may be harmful.

6.2 Characterization of Sources

Sources are entities or actions that release contaminants or other agents into the environment. For chemicals, they may include pipes, stacks, spray nozzles, or sunscreen coated swimmers. For other agents, they may be highways for road kills, seismic surveys for marine noise, or dredges for removal of benthic communities. Sources are your starting point for analyses in predictive assessments. They should have been identified or at least hypothesized in the conceptual model (Chapter 23). However, they should be fully characterized in the analysis of exposure to allow quantification.

Some sources are characterized during the problem formulation based on information provided by an applicant or responsible party. Examples include the rate of release and composition of a proposed effluent and the use patterns of a pesticide or other compound that may be applied to the environment. However, you must supplement these source terms if information is not provided such as the potential composition and rates of accidental or fugitive releases. If the source is a deposited waste at a contaminated site, the source characterization may come from sampling and analysis, perhaps supported by records.

Sometimes proposed source terms are not provided and you must estimate them. For example, the USEPA's assessment of the proposed Pebble Mine was based on a mining plan that did not include estimates of leachates from solid wastes. We modeled those sources from estimated solid waste volumes, precipitation rates, and results of laboratory leaching tests (USEPA 2014a).

Sources associated with agents or activities other than chemical releases may be conceptually analogous to chemical sources but require different quantification. For example, soil erosion from tilled agriculture, construction, and logging can be characterized in terms of the parameters of the universal soil loss equation (Kinnell 2010).

In some cases, such as contaminated sediments or soils, the sources may no longer exist. In such cases, the exposure analysis begins with the contaminated medium. This considerably simplifies the exposure analysis; the assessment and measurement exposures are the same. Contact with the contaminated soil or sediment is the proximate exposure for soil or benthic organisms.

Analysis of sources for causal assessments is often the opposite of analysis for risk assessments. They typically identify the causal proximate exposure for the effect and then work back its source. Hence, for causal assessments, the source is often a result rather than a starting point of the assessment. The process of identifying the source of a causal agent may involve identifying potential sources and routes of transport to the site of exposure, and then determining whether the source and route would result in the magnitude and character of the causal exposure. This process is known as environmental forensics. Sources of exposures may be determined by various techniques depending on the case. They may involve matching the exposure agent to a source by chemical fingerprinting, isotope matching, pathogen genotyping, and similar techniques. It may also involve modeling the routes of transport between potential sources and the place and time of exposure. Finally, the proposed route of transport may be tested by hydrogeochemical tracer studies. For example, chemical fingerprinting and other studies showed that coal-tar-based pavement sealants are the primary source of PAHs in some urban waters (Mahler and Van Metre 2011).

6.3 Analysis of Transport and Fate

Transport and fate models are your tools for translating the source information into relevant concentrations in media or food items at the point of exposure of receptors. They are mathematical simulations of the exposure scenarios. They may be generic models for regulation of chemicals or other agents that are released at many sites. Such models have generic environments and scenarios. They may also be specific models for a real source and receiving environment. For site-specific models, you can adapt a generic model's equations and parameters for the specific location or create a new model. Transport and fate models range from simple dilution models to large and elaborate models that include microbial degradation, photo-oxidation, partitioning within a medium, exchange between media, and biological uptake. An example is the PRZM/EXAMS models that estimate pesticide concentrations in water and sediment from application rates of pesticides to crops. Because the combined models are complex, a shell called EXPRESS was developed to facilitate their application. Environmental agencies have catalogs of models for individual media and multimedia transport and fate modeling. For the USEPA, see the web sites of the Center for Exposure Assessment Modeling and the Support Center for Regulatory Atmospheric Modeling (unless there has been another reorganization). Guidance for modeling in general is also provided by the USEPA (2009).

6.4 Analysis of Proximate Exposures

The fundamental question that you must answer in an exposure analysis is, how do organisms come to contact and take up chemicals and other agents in the environment? Given the answer to that question, ask how do I measure or estimate the exposure in a way that can be used to estimate effects? For most environmental assessments, the answer is a concentration of the agent in water, sediment, soil, air, or food. Usually, it is the total concentration, but it may be a bioavailable concentration. For wildlife and humans, the concentrations usually are converted into doses. The duration of exposure is also relevant, but as in toxicity testing, proximate exposures are usually simply classified as acute or chronic.

6.5 Forms

The form of a chemical in the environment may greatly influence the exposure and effects. The forms are determined by whether the chemicals are dissolved, particulate, gaseous, ionized, or neutral. The forms at the point of exposure are determined by the form released and environmental conditions. For example, the metals speciation submodel used by the USEPA (2007a) in the copper biotic ligand model includes temperature, aqueous cations (Ca^{2+}, Mg^{2+}, Na^{+}, and K^{+}), aqueous anions (Cl^{-} and SO_4^{2-}), sulfide, pH, alkalinity, and dissolved organic carbon. In food items, the chemical form depends on its metabolites, excretion, partitioning to particular tissues, and other processes in the food species. For example, lead partitions to bone, so humans who do not consume bones have less lead exposure than wildlife that consume their prey whole.

For any analysis of exposure, you should consider how the forms of the agent that you measure relate to the purpose of your assessment. An example of failure is provided by a systematic review of the effects of nitrogen and phosphorus on aquatic primary production (Bennett et al. 2018). The review placed a high priority on field experiments as the most useful type of study to define the exposure–response relationship. The desired exposure metrics were total nitrogen and total phosphorus because these are the forms that are used in U.S. water quality criteria and standards (i.e., the assessment exposures). However, of the field experiments, only 3% used total nitrogen and only 5% used total phosphorus. Instead, most used a few readily cycled forms (i.e., measurement exposures). This makes sense from the point of view of the experimentalists who were interested in nutrient chemistry. However, it resulted in frustration for the assessors and the USEPA sponsors of the review because there was no way to extrapolate between those forms. Ultimately, other data were used to derive criteria (USEPA 2021a).

6.6 Modes of Exposure

As discussed above, we estimate assessment exposures which are the ones that are regulated or managed and measurement exposures which are the ones that are modeled or measured in tests or in the field. Another categorization is the modes of exposure: direct contact with contaminated media; ingestion of contaminated food, water, or media; internal exposure; and exposure to physical agents, disturbance and harvesting.

6.6.1 Contact with Media

You may assess exposure by contact with contaminated media in two ways. First, you can consider exposure to the whole medium including the contaminant mixture. This is conceptually simple. Organisms are in the medium and the medium as a whole is the proximate exposure. Second, you can characterize the exposure to each contaminant in a medium as a concentration derived by chemical analysis. To assess the total toxic effects of a medium from concentrations of individual contaminants, you must model the mixture (Chapter 28).

Aquatic exposure assessments are typically based on total or dissolved concentrations of individual contaminants. For aquatic organisms, dissolved concentrations typically approximate bioavailable concentrations and are similar to the exposures in the clean waters of most toxicity tests.

Dissolved concentrations may be measured by filtration. For metals, the dissolved form is commonly used, but the free ionic form is generally considered the toxic form for aquatic organisms. It must be estimated by speciation modeling. Concern about aqueous forms of chemicals may be bypassed by collecting and testing whole aqueous effluents or ambient waters (Section 21.5). Total aqueous concentrations are used in most human health assessments, so those concentrations may be the only aqueous data available to assessors.

Sediment-dwelling organisms may be assumed to be exposed to contaminants in the whole sediment or to only the sediment interstitial (pore) water. Whole sediment concentrations are obtained by extraction and analysis of the extracts. Pore water concentrations are generally accepted to be a better predictor of toxicity than whole sediment concentrations. Pore water may be obtained by filtration and analyzed. It is commonly assumed that organic compounds are in equilibrium between the water and sediment organic matter. Hence, equilibrium partitioning coefficients can be used to convert total concentrations to pore water concentrations. This approach has been used to develop sediment benchmarks (USEPA 2002b, c, e). A similar approach for metals involves complexation by acid volatile sulfides (USEPA 2002d). It is more difficult than partitioning of organics and less generally used. Sediment is chemically and physically complex and is highly variable in vertical and horizontal space. It varies less over time than water but can be drastically changed by storm events that scour or by stir sediments.

Soil contact exposures are determined for endpoint organisms that are in the soil and contact associated contaminants: plants, invertebrates, and microbes. Most commonly, total extractions are used to estimate field soil concentrations and to characterize contaminated field soils. Concentrations in aqueous phases of soils may be approximated with aqueous extractions or partitioning coefficients.

6.6.2 Ingestion of Food, Water, and Soil

You can use the same basic model to estimate the ingestion exposures of wildlife and humans to food, drinking water, and soil. In the simplest form, exposures are products of ingestion rates and contaminant concentrations in the ingested media ($E = i\ c$). Ingestion rates expressed as milligrams of solids per kilogram of organism per day or as liters of liquids per kilogram per day. Concentrations are milligrams of contaminant per kilogram of food or soil or milligrams per liter of water or other liquids such as liquid wastes. This basic equation may be expanded by summing exposures across media or across contaminants with a common mode of action (Chapter 28).

Human ingestion rates of various foods, soils, and liquids are well documented in Agency reports (USEPA 2008, 2011a, 2019b). Concentrations of contaminants are provided by various monitoring programs. For example, pesticide concentrations in U.S. foods are monitored by the U.S. Department of Agriculture's Pesticide Data Program and used by the USEPA to estimate human exposures.

Wildlife ingestion exposures are not as well studied or documented. The basic equations are the same as for human health assessments, but because of the many species and food sources, the parameters are not nearly as well defined by measurements. Instead, models based mainly on the size of the organisms, but sometimes on their diets, taxonomic groups, or other factors are used to estimate ingestion rates of foods and water. What is available from the USEPA, ORNL, and others in North America has been collected into the *EPA EcoBox Tools – Exposure Factors* and – *Exposure Pathways*.

The ingestion of soil is not an obvious risk issue like food and water, but it can be important. Terrestrial vertebrates inadvertently consume soil associated with food such as soil within and on earthworms, adhering to roots, or on the fur of mammalian prey. They also incidentally consume soil when grooming. They deliberately consume soil that is high in sodium or other mineral nutrients. This may include soils contaminated with wastes that include nutrient minerals like the coal ash on the Oak Ridge Reservation or have attractive properties like the sweet taste of ethylene glycol (Sample and Suter 2002). Soil ingestion rates for wildlife are compiled by Suter et al. (2000) and USEPA (2005). However, it has been suggested that site-specific soil ingestion be determined for endpoint wildlife at contaminated sites (Sample et al. 2013). Humans may consume soil deliberately because it is thought to have health or spiritual benefits, incidentally on food, or by children playing in the dirt and sticking their fingers in their mouths (USEPA 2011a, 2008).

6.6.3 Internal Exposures

You can confirm that an exposure has occurred by using measurements of concentrations of the agent in whole organisms, organs, or tissues. This information is particularly useful for causal assessments because it confirms that exposure has occurred. However, it can also support risk assessments. It is difficult to use internal concentrations for exposures in risk characterization because few exposure–response relationships use internal concentrations. Also, they are difficult to interpret because some tissues sequester chemicals. For example, lipid-soluble organic chemicals are found in fat stores where they are effectively inert, until they are released by fat metabolism or lactation.

Internal exposures are also used to assess risks to the consumers of the analyzed organisms. The appropriate analysis depends on whether the whole organism or certain parts are consumed. For example, for human health assessments, typically only the fish filet is analyzed. Piscivorous wildlife often consume the whole fish, but in some circumstances certain parts are preferred. For example, when salmon are abundant, bears will eat only the gravid ovaries and other fatty parts. The bears are after the calories, but they also get lipophilic chemicals.

If internal concentrations are not determined by analysis, they may be estimated. The easiest and most common method in ecological assessments is bioconcentration factors (concentration in an organism/concentration in an abiotic medium) or bioaccumulation factors (the same quotient but including food as well as direct uptake). The alternative approach is mechanistic modeling using toxicokinetic models, which are found mostly in human health assessments.

6.6.4 Exposure to Physical Agents

Exposure for projects typically involves the loss of ecosystems. Examples include construction or mining in terrestrial ecosystems or conversion to agriculture; draining or filling wetlands; and channelizing, damming, or burying streams in culverts. The units of exposure are simply linear meters of streams or square meters of lands, wetlands, lakes, or marine waters.

Harvesting is similar in that it involves destruction and removal. Examples include fisheries, logging, and hunting. Some processes are analogous to harvesting such as entrainment and impingement of fish in power plant cooling systems and hydroelectric systems. Exposure is expressed as the number of organisms removed, usually partitioned by species and age or size class.

6.7 Presenting an Exposure Characterization

An exposure characterization (aka exposure profile) is a summary of the exposure analysis results. For risk assessments, it should briefly describe how the analysis is performed and tabulate the results that will be used with the results of analysis of effects (Chapter 7) to perform the risk characterization (Chapter 8). Analyses of exposure must also be characterized for other types of assessments in an appropriate form. For example, in an outcome assessment, exposure is characterized to determine whether remediation was successful.

7

Analysis of Effects

The analysis of effects for risk and other predictive assessments has two components: hazard identification and exposure–response analysis.

7.1 Hazard Identification

The hazard of an agent is its intrinsic capacity to cause an effect. The clearest hazard is death, particularly acute mortality (Box 7.1). The conventional ecological chronic hazards are reduced survival, growth, and reproduction, all of which may contribute to reduced population size or growth. In human health assessments, cancer is treated as a distinct hazard because it is exceptionally dreaded. Noncancer human health hazards are diseases such as asthma, impairments such as IQ decrement, and physical abnormalities such as developmental defects. The hazards of concern are those that correspond to an assessment endpoint (Chapter 14) or that can be extrapolated to an assessment endpoint (Chapter 26).

The capacity to cause an effect is an inherent attribute of an agent (selenium causes deformities in fish), but there is no hazard without a possible source and route of exposure. For example, selenium leaches from coal mine wastes into streams. Selenium on Mars is still a teratogen, but it is not a hazard.

In many cases, you identify the hazard associated with an agent during the problem formulation (Chapter 4). However, hazard identification may be revisited or performed entirely in the analysis phase. That is because the hazard may be unknown or uncertain until assessment-specific information has been generated and data analyses have been performed. If hazards are identified in the analysis of effects, you must consider whether you need to create or expand the list of assessment endpoints (Chapter 14). Assessment endpoints are specific attributes of specific entities that may be susceptible to the hazard. For example, if a hazard of a chemical in a waste is that it is a carcinogen, a human health assessment endpoint might be the frequency of cancers in individuals residing within a kilometer of the contaminated site.

DOI: 10.1201/9781003156307-8

BOX 7.1 ACUTE AND CHRONIC

Acute and chronic are important terms in environmental toxicity, but you may find them confusing because they are used ambiguously. In toxicology and assessments, they refer to temporal durations of exposures and associated effects. An acute exposure is brief, as in applications of pesticides, spills of chemicals, failures of wastewater treatment plants, or passage through hydroelectric turbines. In mammalian toxicology, acute has been defined as an exposure of less than 10% of an organism's life-span. In aquatic toxicology, its limit is typically 2–4 days. Chronic exposures are any that are longer than acute exposures and they may be either continuous or recurring (e.g., occurring annually during a low flow season).

Acute effects result from acute exposures. However, in ecological assessments, acute effects include only mortality or equivalent effects such as immobilization. Other effects of short ecological exposures are not considered because they are negligible (e.g., reduced growth for a day) or because recovery is expected.

Chronic effects are those occurring for a long time either because the causal exposure is chronic or because there is no recovery from an acute exposure. The standard chronic ecotoxicological effects are mortality, fecundity, and growth. For humans, a chronic effect is variously defined as continuing for anywhere from 3 months to a year or more.

There are intermediate cases. For example, a 7-day survival and growth test of larval fish is usually referred to as subchronic.

7.2 Exposure–Response Relationships

The heart of any assessment is the causal relationships between the exposure and the resulting effects. Most assessments require that you quantify the relationship. You may express exposure–response relationships in several ways.

The simplest relationship is a threshold which may be a designated benchmark value (Chapter 22) or a value that you develop for a case such as "no fish were collected where contaminant concentrations exceeded 10 mg/l." A threshold value is, in effect, a single-point exposure–response relationship. Your threshold values should be based on the environmental or policy significance of effects, not statistical significance (Chapter 24). Commonly, 5% or 10% effect level is used to derive benchmarks and case-specific threshold values.

Another common approach is statistical modeling. The most common type is a function such as the log-logistic fit to toxicity test data (Figure 7.1). These

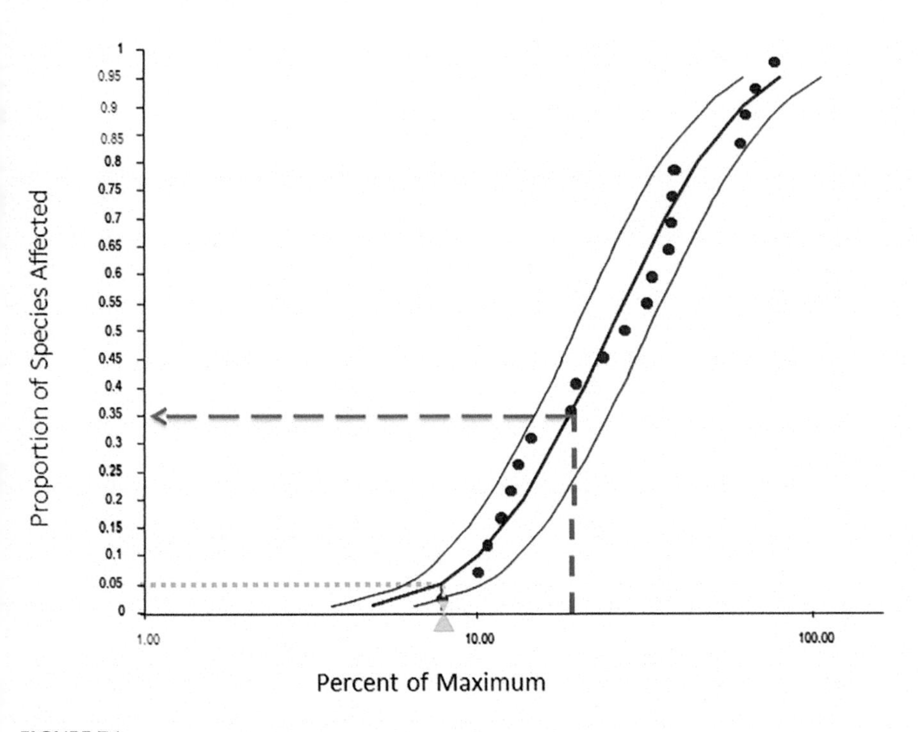

FIGURE 7.1
A generic exposure–response model from a toxicity test with 95% confidence limits. The dark gray dashed line illustrates using the model to estimate the level of effect. The light gray short-dashed line illustrates using the model to estimate the level of exposure (at the triangle) that would cause a certain effect.

are typically concentration– or dose–response relationships that estimate the frequency of mortality or other dichotomous effects as exposure increases. They may also model magnitudes of continuous variables such as growth or count variables such as fecundity. The most common exposure variables in these models are concentration and dose, but time is sometimes very useful. Duration–response relationships can be used to decide, for example, how long an untreated effluent may be released during an upgrade of a water treatment plant. Multivariate regression can be used for exposures to multiple agents (Chapter 28). Environmental agencies like the USEPA have software and guidance for statistical modeling of exposure–response relationships (https://www.epa.gov/bmds) as do statistical software packages like R.

Most statistically derived exposure–response relationships that are based on distributions of data concerning effects on organism-level attributes test endpoints are responses of organisms. However, an important alternative is species sensitivity distributions (SSDs—Chapter 27). Rather than representing a population as a distribution of organisms, SSDs represent biotic communities as distributions of species.

Finally, you may create an exposure–response relationship by mathematically modeling the processes that characterize causal relationships. The most commonly used process models in ecological assessments are the population demographic models that are used to assess fish and wildlife harvesting programs and sometimes risks from pollutants (Forbes, Calow et al. 2011). The most practical examples are cases in which the only effect is mortality, as in assessments of harvesting programs. Harvested species are modeled because they are important and they should have existing demographic data. For example, to assess effects on a game fish population of entrainment in the cooling system of a power plant, you could use a demographic model of the population of concern (Barnthouse 2013). By combining age or life-stage specific mortality rates from entrainment with the background rates, you could estimate changes in population size. For a chemical that affects growth and reproduction, decrements in both of those rates may be used in the model. Organism-level effects can be modeled in terms of the toxicokinetic processes that translate uptake into concentrations at which proximate effects are initiated and toxicodynamic processes that translate the initiating event into an endpoint effect. Finally, ecosystem models may be used to derive ecosystem-level endpoints from exposure–response relationships for species or trophic groups. These models are rarely used in the management of chemicals or other agents in the environment, but they are available from the USEPA and some other agencies (https://www.epa.gov/ceam/aquatox).

Exposure–response modeling for human health is largely limited to fitting statistical functions to mammalian test data. In many fewer cases, epidemiological studies can relate exposure levels to effects. An example is Herbert Needleman's pioneering studies relating lead levels in children to cognitive and emotional effects. For an inspiring account of Needleman's work and its use in the USEPA lead assessment, see Denworth (2008). Suborganismal mechanistic studies are becoming more common in human health assessments, but currently they are used more to estimate parameters of internal exposures or to identify hazards than to estimate exposure–response relationships for endpoint effects.

If, as is often the case, you derive an exposure–response relationship from data for a species, effect, duration, and conditions that are not the ones needed for the assessment endpoint, you should extrapolate (Chapter 26). For a simple example, to derive a benchmark when you have a rainbow trout LC50—a point exposure–response relationship for death of half the fish—you might divide it by a factor of 2 to increase the proportion of fish species that would be protected from acute lethality.

7.3 Presenting an Effects Characterization

The effects characterization (aka effects profile) is a summary of the effects analysis results. For risk assessments, it should briefly explain how the analysis is performed, describe the hazard, and present the exposure–response relationships for each assessment endpoint. The characterization should be sufficient to be used with the results of analysis of exposure (Chapter 6) to perform the risk characterization (Chapter 8).

8

Risk Characterization and Other Syntheses

All scientific work is incomplete—whether it be observational or experimental. All scientific work is liable to be upset or modified by advancing knowledge. That does not confer upon us a freedom to ignore the knowledge we already have or postpone the action that it appears to be demanded at a given time.

—A.B. Hill (1965)

Your risk characterization or any other synthesis phase of an assessment should have two major components. The first is an analysis that synthesizes the exposure and effects and explains the methods used to derive them. You should communicate those results in a way that clarifies the importance and adversity of estimated effects, the evidence supporting the estimates, and the potential implications (USEPA 1998, USEPA 2014c). The second component is the context of those results which explains their implications. By analogy to a scientific paper, in the characterization, you are creating a combined results and discussion section for the assessment. A decision-maker should be able to skip the problem formulation and analyses, read only your synthesis, and come away with sufficient information to inform the decision.

This chapter focuses on risk characterization. However, the concepts apply to syntheses of other types of environmental assessments (Section 8.5).

8.1 Components of Results

A risk characterization derives and presents the key results of the assessment.

Hazard—A hazard is a potential effect of exposure to some agent or action. The hazards to be addressed in an assessment are defined during the problem formulation. However, during the effects analysis and risk characterization, they may be confirmed, modified, added to, or reduced.

Integrated results—The integration combines the results of the analyses of exposure and effects to estimate the magnitude or frequency of effects. This can be done in multiple ways (Section 8.4). Ideally, in risk assessments, the exposure–response relationship is solved for the estimated exposure to derive an estimate of effects. However, if the analysis of effects uses an

DOI: 10.1201/9781003156307-9

existing benchmark value (Chapter 22), the exposure estimate is divided by the benchmark value to derive an assessment quotient. If the assessment derives a benchmark, an exposure–response relationship is solved for the threshold effect. Your description of a risk estimate should indicate what you are assessing (e.g., individual, population, ecosystem) and include the high end and central tendency estimated effects.

Confidence—Determine the degree of confidence in the estimated effects for each hazard and the range of plausible levels of effects (Chapter 17). If weight of evidence is used in the inferences, summarize the confidence provided by the body of evidence (Chapters 24, 25).

Uncertainties—Identify the remaining uncertainties and present their implications (Chapter 17).

Methods—Explain how the hazards, integrated results, confidence, and uncertainty were derived.

8.2 Components of the Discussion

Your characterization should provide context and explain the implications of the analytical results. The issues to be discussed and the nature of the discussion depend on the type of assessment.

Scope—The scope of the assessment includes defining the spatial and temporal extent, the range of actions, agents, media, environments, and biota considered.

Strengths and weaknesses—Explain what aspects of your assessment give you confidence in your results and what limitations of the assessment result in weaknesses that may limit applicability. In particular, describe important assumptions and extrapolations. If an explicit weighing of evidence was not part of the analyses and synthesis, at least provide judgments concerning the relevance, strength, and reliability of the body of evidence used.

Reasonable alternative interpretations—Other assessments by other organizations might have come to other conclusions or might draw alternative conclusions based on different information, scientific opinions, or policies. You should make decision-makers aware of these alternatives and why your results are appropriate.

Importance—In what sense are the results adverse and severe, if they are? This is particularly important for ecological assessments which may have unfamiliar endpoints. Consider public opinion, laws and regulations, precedents, and ecosystem services. The importance of human health effects is more obvious, but may require some explanation if the endpoint is not death or equivalent severe and overt effect.

Implications for the decision—Explain how the results of the risk characterization inform the decisions to be made. This does not include stating what the decision should be.

8.3 Manner of Conducting Risk Characterizations

The USEPA's risk characterization policy calls for conducting risk characterizations in a manner that is consistent with the following principles known as TCCR (Science Policy Council 2000).

Transparency—The characterization should fully and explicitly disclose the risk assessment methods, default assumptions, logic, rationale, extrapolations, uncertainties, and overall strength of each step in the assessment.

Clarity—The products from the risk assessment should be readily understood by readers inside and outside of the risk assessment process. Documents should be concise, free of jargon, and should use understandable tables, graphs, and equations as needed.

Consistency—The risk assessment should be internally consistent and should be conducted and presented in a manner which is consistent with relevant policies and precedents.

Reasonableness—The risk assessment should be based on sound judgment, with methods and assumptions consistent with the current state-of-the-science and conveyed in a manner that is complete, balanced, and informative.

8.4 Expressing Results

The results of environmental assessments can take multiple forms. Three considerations are relevant: what expressions can reasonably be derived given available information and resources, what expressions make sense for the type of assessment and the particular case, and what expressions are acceptable given precedents and preferences of decision-makers?

8.4.1 Hazards and Other Qualitative Results

Some assessments answer qualitative questions such as what hazard is posed by this agent, what caused this observed impairment, or is this chemical a persistent, bioaccumulative, and toxic substance? The results should be clearly and simply expressed as in, the critical hazard posed by dietary carbofuran is neurotoxic effects on the central nervous systems of children. The result of a causal assessment could be, ammonia is the cause of observed toxicity and the sewage treatment plant is the source.

8.4.2 Quotients

Assessments of the acceptability of an exposure are usually expressed as assessment quotients (also known as risk or hazard quotients). These are exposure levels divided by a benchmark value:

$$Q = E/B$$

E is a measured or estimated concentration, dose, or an equivalent value such as temperature. *B* is the benchmark, which is a standard, criterion, guidance value, screening benchmark, or equivalent value that is a threshold for acceptability (Chapter 22). If *Q* is greater than 1, the exposure is not acceptable with some implication for action. If *B* is a regulatory standard, a fine may be levied or an effluent may require treatment. If it is a screening benchmark, additional measurement, testing, or assessment may be required.

Quotients may be conservative because conservatism may be incorporated in *E* or *B*. The parameter *E* may be the highest measured value or may be modeled using conservative assumptions or scenarios, such as atmospheric exposure estimated at a facility fence line. The parameter *B* may be the exposure level at which the most sensitive response of the most sensitive species is estimated to occur. In EPA human health risk assessments, *B* is a test endpoint divided by factors of 10 for each extrapolation: from animal to human, from typical human to sensitive human, and possibly other factors (Chapter 26).

Quotients are often disparaged as simplistic. However, considerable measurement, testing, and modeling effort may go into derivation of *E* and *B*. That often includes adjustments for uncertainties to assure protectiveness. More importantly, the clarity of *Q* makes the assessment understandable to decision-makers and the public. The consistent use of a bright line ($Q=1$) supports a regulator's contention that they are being transparent and not arbitrary or capricious. They are equivalent to enforcing highway speed limits, where the speed limit is *B*, the radar gun's reading is *E*, and $Q>1$ is a violation.

8.4.3 Magnitudes of Effects

The most obvious expression of ecological or human effects is a quantitative estimate of the magnitude, frequency, or duration of effects. This can be done in multiple ways. Conventionally in risk assessments, the exposure–response relationship is solved for the estimated exposure to derive an estimate of the magnitude or frequency of effects. Sometimes an exposure estimate is sufficient. For example, the Bristol Bay assessment estimated that wetlands would be filled or excavated in 10.2 and 17.3 km^2 of the minimum and maximum mine footprints, respectively. That is a sufficient result, because wetlands are legally protected and a wetland that is filled or engulfed by a mine pit is not a wetland. Quantitative expressions of confidence in an estimated effect are confidence intervals or, as in the Bristol Bay example, intervals based on the range of effects for a range of scenarios or assumptions.

8.4.4 Probabilities

Probability is a source of contention in risk assessment (Chapter 18). Many risk assessors, including me at one time, believe that risks are probabilities and the results of risk assessments should be probabilities. However, experience has taught me that, with few exceptions, decision-makers do not want to be told that there is a probability of 0.67 that an effect will occur.

8.4.5 Relative Risks and Odds

If your results are frequencies of exposures or effects, alternatives to simply reporting the relative frequencies or converting them to probabilities are relative risks and odds ratios. Both of these ratios express the strength of association between exposure and effects. These metrics are uncommon in environmental risk assessments, but are more common in other assessments such as risks of accidents and failures and for results of medical treatments. You may find them to be useful and you should understand them when you see them. Table 8.1 and the accompanying text explain the relationship of frequencies to these other metrics.

The natural frequency and proportion (probability) of effects given exposure are 71/116 and 0.61.

The natural frequency and proportion (probability) of effects given no exposure are 15/155 and 0.097.

The relative risk is the quotient of the natural frequency of effects given exposure divided by the natural frequency of effects given no exposure $(a/a+b)/(c/c+d)=(71/116)/(15/155)=6.3$.

TABLE 8.1

A Table of Frequencies of Effects and No Effects and of Exposure and No Exposure

		Effect		
		Yes	**No**	**Totals**
Exposure	Yes	a=71	b=45	116
	No	c=15	d=140	155
Totals		86	185	271

The numbers are arbitrary but plausible for environmental studies. Letters are used in the formulas.

Note that proportions are 0–1 scaled frequencies, and probabilities applied to frequencies like this are just proportions. Probabilities are scaled 0–1 to match proportions (Clayton 2021). However, proportions are just numerical descriptions, whereas probabilities are variously defined as chance associated with random devices (e.g., dice and flipped coins), frequencies, or degrees of belief (Chapter 18).

The odds of interest are relative frequencies of exposure given frequencies of effects. That is, it is answering a question about causation of observed effects. The odds ratio tells you how much higher the odds of exposure are among affected organisms than among the unaffected.

The odds of effect versus no effect in the exposed group are 71:45 or 1.58.

The odds of effect versus no effect in the unexposed group are 15:140 or 0.107.

The odds of effect in the exposed group relative to the odds of effect in the unexposed group are the odds ratio $(a/b)/(c/d) = (71/45)/(15/140) = 14.7$.

These odds ratios and relative risks are for experimental studies such as toxicity tests or epidemiological and ecoepidemiological cohort studies. For example, studies comparing the invertebrates in streams draining valley fills to those draining unfilled valleys (i.e., exposed and unexposed cohorts). If you are analyzing a case-control study such as comparing a set of impaired streams (cases) to unimpaired streams (controls) with respect to frequency of valley fills, you cannot use relative risk because the number of affected streams is set by the investigators. Using Table 8.1 again, the odds ratio for case-control data is the odds that a case (an affected entity) was exposed relative to the odds that a control (an unaffected entity) was exposed $(a/c)/(b/d) = (71/15)/(45/140) = 14.7$. Surprise, the result is the same as for cohort studies. This is a source of confusion because some definitions of odds ratios give one formula and some give the other without explanation.

Relative risk is reasonably intuitive: $RR=1$ indicates no effect of exposure, $RR>1$ indicates increased effects with exposure, and $RR<1$ indicates decreased effects with exposure. Odds ratios are scaled equivalently, so that larger deviations from 1 in odds ratios indicate stronger positive or negative associations. However, odds ratios are often confusing, so even if they make sense to you, odds are they will take some explanation when you present them. In this hypothetical case, the association of exposure and effects is strong by either metric, but as is usually the case, odds ratios are larger than the relative risk.

8.4.6 Significance of Results

Results of an assessment should be summarized in a way that conveys their significance. Results are significant when they are sufficiently important to influence a decision. It has four components:

- First, the nature of the effect is more or less important. Does it involve a species or an ecosystem with designated conservation status, public interest, or economic importance? This should have been considered when defining the assessment endpoints (Chapter 14), but should be reconsidered during this final stage and conveyed in the results.
- Second, the magnitude of effects can and should influence the significance of results. This includes the spatial and temporal extent of effects and the number or proportion of organisms, populations, or ecosystems affected (Section 8.4.3).
- Third, the confidence in the results and the importance of any residual uncertainties (Chapter 17). Results are more significant if you are confident in them and if residual uncertainties are unlikely to make you unsure.
- Fourth, the relationship of the causal agents to mandates for action and means to act effectively (Chapter 10). Results are more significant if someone will be compelled to do something in response to them.

8.5 Synthesis in Other Types of Assessments

Risk characterization guidance is applicable to risk assessments, but all types of environmental assessment require some sort of synthesis of the evidence and analyses to generate the results. For example, the results of an assessment to derive a water quality benchmark should include the quality expressed, the numerical result and units, quantitative confidence, and confidence in the evidence (Chapter 22). Synthetic results from a condition assessment would include statistical summaries of ambient concentrations related to any relevant benchmark values, summaries of biological survey results, and comparisons to data from reference sites. Causal assessments have a characterization of causes step after analysis of the evidence. It describes the cause and confidence in its determination. The TCCR principles are applicable to the synthesis step of all types of assessments.

9

Communication

No wise man has the power to reason away what a fool believes.
—*Kenny Logins and Michael McDonald,* What a Fool Believes

Communication is the process in which assessors exchange information and opinions with decision-makers, stakeholders, and the public. In environmental assessment, communication occurs primarily before the analysis and writing begin and after the documents are completed. For example, in site-specific assessments, pre-assessment discussions with stakeholders or the public may reveal important exposure pathways to include in the assessment such as subsistence or commercial fishing. After the draft assessment report is completed, it may become the object of public, stakeholder, and peer review which may result in extensive edits and revisions. These reviews may raise new issues that must be assessed, errors that must be corrected, and ambiguities that must be clarified. The final assessment report is communicated to the decision-maker and may be released to the public.

The primary assessment document is a technical report that describes the project, impairment, or agent being assessed; the potential actions to be taken; the scope of the assessment; the analyses performed; and the characterization of the results in terms of conditions, risks, causes, or impacts. It may encompass ecology, human health, or both. These documents are often lengthy and difficult for a nontechnically trained reader. However, if technical details are not provided, peer reviewers and reviewers from responsible parties, proponents, and opponents will not be able to independently judge your work. It may be helpful to write an extended abstract for each section to cover the topic for an educated audience and to let an environmental scientist or engineer know what is coming. Put lengthy material that is important for documentation, but not for understanding, in appendices so that they do not bog down the reader.

The process of written communication may be well served by separate documents designed for particular audiences. The most common example of this is the executive summary. Unfortunately, these are often given little thought, and I have even seen cases in which the introductory section of an assessment was copied and pasted as the executive summary. These summaries are not just for executives. For major assessments, they may be stand-alone documents that are also for the public and for nontechnical stakeholders. They

DOI: 10.1201/9781003156307-10

should be readable and provide maps, graphics, and photos of the site and perhaps endpoint species. In other words, they should be visually appealing and informative to people who respond to different communication modalities. See for example, the executive summary for the Bristol Bay Watershed assessment (USEPA 2014b). However, few members of the public or stakeholders are engaged with most assessments and many assessments do not have the time or resources for high-quality summary documents. In those cases, a simple executive summary written for the decision-maker can be sufficient.

Oral communication prior to the assessment may be relatively informal. It elicits information and opinions from decision-makers and stakeholders concerning plans for the assessment. You should guide the conversation to ensure that all topics are covered. When the assessment is complete, your oral communications switch to more formal presentations of the results. You suit the presentation to the audience in terms of level of technical information and style of presentation. Speak directly to a manager being briefed and be prepared to be interrupted and questioned. When engaging a public audience, look at them and "read the room." Follow the usual advice about talks, practice it, make sure it fits the time you will have available, project to the back of the room, etc.

9.1 Communication and Neutrality

You should remember that your job when communicating assessment results is to lay out your findings in a neutral manner, including your confidence and uncertainty. Just as you suppress your potentially biased opinions when performing an assessment (Chapter 20), you should also suppress them when communicating. This becomes particularly important when you are communicating with advocates for or against the project or product being assessed. Social psychologists have shown that it is difficult to change people's established opinions. In my experience, you lose your neutral stance if you try to correct or convince them, even when the facts are on your side. Keep the Logins and McDonald lyric in mind, and listen respectfully to even nonsensical opinions.

9.2 Communication with Decision-Makers

The primary function of environmental assessment is to provide scientific information to inform a decision-making process. Hence, communicating with decision-makers is a core function of assessors. However, our relationships with decision-makers are complex. On the one hand, there is a concern that if there is too much communication with decision-makers, assessors will be pushed to bias their work to favor the decision-maker's preferred outcome. This was a major concern in the early days of risk assessment in the U.S. which led to the identification of a distinct planning stage in which decision-makers would express their needs, and then they would step aside until the assessment was complete (NRC 1983). As experience was gained, we found that a bigger problem was assessments that did not meet the decision-maker's information needs. Because of the many choices such as identifying scope and endpoints that were made in the problem formulation after the planning stage, the decision-maker should be available. Now, in USEPA practice, the decision-maker may be involved during any stage of an assessment.

Assessors may play two different roles when communicating with decision-makers, technical consultant and advisor (Suter and Cormier 2012). Acting as a technical consultant means doing all of the things that assessors are charged to do, resulting in a report and probably a briefing. Assessors do not routinely act as advisors, but you should be prepared to provide requested advice. Advice may include explaining precedents for the decision in response to questions such as "has anyone done a remediation based on insect diversity?" Some questions may seem silly but should be taken seriously. I have been asked, "why should I care about bugs in the mud?" In the extreme, a decision-maker might ask what you think should be done. In such cases, you should not directly answer. Rather, refer back to the assessment and explain how its conclusions seem to you to support one action more than another. In other words, provide the requested opinion but to keep as close to the science as possible.

You should know who the decision-maker is. In the USEPA, routine decisions like permit renewals are made by a career professional. More important or controversial decisions are made by political appointees such as regional administrators and assistant administrators for the offices of air, water, waste, and toxics. In very prominent and controversial decisions such as the permit for the Pebble Mine in Bristol Bay, Alaska, the decision is made by the administrator or even the president. During the course of an assessment, the decision-maker may change. The regulatory process for the Pebble Mine was overseen by five USEPA administrators appointed by three presidents. If possible, you should learn the decision-maker's background. How familiar are they with the science and with the assessment and regulatory process? Some are remarkably ignorant and need a lot of explanation; others are scientifically trained or have extensive relevant experience.

In some cases, there are intermediate decision-makers to communicate with. A high-level decision-maker may have support staff who prepare a decision document. If possible, you should engage with those people to ensure that they correctly interpret the assessment. In the case of the Bristol Bay assessment, the Office of Water staff misinterpreted the assessments of toxic effects and habitat destruction and assumed that the habitat losses from different causes could be added. I had to explain that during the mine development, a particular stream reach might have toxic levels of copper, but later it would be physically destroyed as the mine expanded.

9.3 Communication with the Public

Public meetings on environmental assessments typically begin with a presentation by assessors, management, or both. The presentation is low on technical details, but it must convey the conclusions and the science-based process behind them. The presentation is followed by statements from the public and sometimes by stakeholders. For example, at a public meeting for the assessment of mining in the Bristol Bay Watershed, the mining company CEO sat with the public but read a prepared statement denying the results of the assessment. Because the purpose of public meetings is to collect public opinions, in general you have no opportunity to request clarification of comments or to respond to comments. Sometimes a moderator is engaged to act as a buffer.

Public meetings can be quite emotional depending on the issue. The Anchorage public meeting on the Bristol Bay assessment had an overflow crowd in the largest auditorium in the city and emotions were high on both sides. There were even credible death threats against the USEPA employees, so Federal Protective Service agents were stationed in the auditorium. That is an extreme example but even in small local public meetings can be emotional. Twice I have seen people holding up babies to emphasize their opposition to a project. You may be sympathetic to a Yupik elder who states that if a mine is approved there will be no future for his little granddaughter or a woman who believes that a synthetic fuels plant will poison her baby, but your role is simply to take note of any information that might contribute to the assessment. That, at least, is the practice in the U.S. Federal Government. In other contexts, public engagement may be more than soliciting and noting comments.

If you are Agency or corporate staff, you should listen attentively, take notes, and answer only specific technical or process questions. Some participants have information to share but most are there to express their support or opposition to the project or action being assessed. Do not correct speakers and do not even ask for clarification when a comment does not make sense. You may be perceived as challenging and even hostile. Do not try to get speakers on your side or assure them that you are on their side. Be a good respectful listener and afterward compile the information obtained and identify frequent issues.

9.4 Communication with Stakeholders

Communication with stakeholders is highly variable depending on the context. For environmental impact assessments, requests for pesticide permits, and some other assessment types, the applicant or responsible party provides an environmental report. A consultant or industry employee, may be the author of such reports and may represent the contractee-stakeholder in meetings with government agencies. If you are with a government agency, you may have opportunities to request clarification or additional information. You may also consult with other agencies, tribal governments, and local governments and they may contribute to the problem formulation. After the assessment is released, at least in the U.S., environmental and business advocacy groups are treated as part of the public. They provide public comments and attend public meetings.

9.5 Communication with Peer Reviewers

We seldom consider that peer review is a communication process. If the assessment is reviewed independently like a journal submission, you will not have an opportunity to directly communicate with reviewers. If the assessment is high profile or particularly complex, a peer review workshop may be held. Communication in those workshops is relatively easy because the peers should be able to understand the process, methods, and results (but not always). If you use unfamiliar and difficult methods, you will lose some or all reviewers. Even the format of the assessment may need to be familiar. Although most ecological and human health risk assessments are organized in terms of a risk assessment framework, that is not necessarily the best format for clear communication. These frameworks were designed to direct the process of assessment and not as a format for writing assessments. However, a peer reviewer insisted that our use of a format other than the ecological risk assessment framework proved that nobody on the assessment team knew anything about ecological risk assessment. Many hours were spent reorganizing the assessment to follow the USEPA framework because a peer reviewer could not deal with a clear but unfamiliar format.

Besides communicating with reviewers through the assessment documents and workshop presentations, you may have the opportunity to help select the questions for the reviewers. The questions should direct the reviewers to the important issues and thereby avoid unhelpful or irrelevant comments.

9.6 Contractors and Communication

This chapter has focused on communication by assessors who are presumed to be part of the organization that needs and is paying for the assessment. The situation is somewhat different for contractors. An in-house assessor is part of the organization, aware of the culture including the environmental attitude of the decision-makers and should have some credibility within the organization. As a contractor, you do not have those advantages. You may communicate with an environmental scientist or engineer who is responsible for the assessment and never directly communicate with the decision-makers or stakeholders. In general, you are somewhat distanced from the setting of goals and scoping of an assessment. As a result, you must focus on exactly what is wanted from them and when. During a peer review workshop, you may be asked to present a particular part of the assessment that you prepared or answer questions concerning work that you did.

In-house assessors, in general, acquire the reputation of their employer, and contractors acquire the reputations of the companies or agencies that you choose to work for. However, if you are a contractor, you will develop your reputation and sell your services based primarily on the quality of your work. Some contractors are leaders in environmental assessment and are conspicuous participants in professional societies. One communication problem for contractors is trying too hard to please their client and perhaps thereby getting more work but also endangering their reputation. In one case, a USEPA colleague and I were asked by a data analysis contractor which result we wanted—not good. Another issue is deciding how to do an assessment task. Unless the contractee has clearly specified the methods, you have a choice between routine default methods and advanced novel methods. The latter would give you a chance to gain attention, build a reputation by publishing and presenting those novel methods, and potentially improve the assessment results. On the other hand, your contractee may specify that you cannot publish any of the work performed under contract. It is important to work out expectations and constraints in consultation with your contractee.

9.7 Advocates and Communication

You may be an assessor for an advocacy organization. These include environmental advocacy organizations such as the Natural Resources Defense Council, Sierra Club, Environmental Defense Fund, Climate Reality Project, and European Environmental Bureau. They also include industry advocacy groups such as the National Mining Association and American Chemistry Council that oppose regulation of their industries and generic anti-regulation organizations such as The Heartland Institute which has argued that cigarettes are safe, that Rachel Carson killed more people than Hitler, and that climate change is not anthropogenic and not harmful. Finally, some organizations are not advocates for a position, but rather are advocates for their unbiased scientific findings. The obvious example is the Intergovernmental Panel on Climate Change (see below).

Many advocates believe that, if the public only knew what I know, they would believe what I believe. However, fact-filled assessments seldom change established opinions. People engage in cultural cognition and take positions that reinforce their cultural identity. Similarly, people engage in motivated reasoning, accepting only conclusions that protect their beliefs and interests. To change minds, you must make you message nonthreatening by presenting information in a values-neutral context. Also, many scientists have a warmth problem and therefore a trust problem.

9.8 Communication Guidance and Models

Most of us have little training in written, oral, or graphic communication. If anything, we are taught a very technical style suitable, at best, for journals or our colleagues. That is not suitable for the public or for decision-makers who are often lawyers, businessmen, or public policy experts. Because of the diversity and complexity of information conveyed in assessments, even our colleagues and peers may be confused by our attempts at communication. Much of the problem is the curse of knowledge. We know our analyses and conclusions better than anyone else and, as a result, we have difficulty remembering to include essential information that our audience does not know. Some of the problem is we do not take the time to improve our communication and instead submit our first draft or limit our graphics to familiar PowerPoint options. You should use the usual writing advice such as avoid acronyms, jargon, long sentences, and grammatical errors. Write in paragraphs, each of which is devoted to one topic introduced in the first sentence. Use a style guide such as Pinker (2014).

There are many books, guidance documents, and web resources that can help you become a better communicator. Organizations that work to make scientists more persuasive include COMPASS (https://www.compassscicomm.org/) and the Alan Alda Center for Communicating Science (https://www.aldacenter.org/). Also, environmental organizations such as the USEPA have resources for communication directed to their communication needs (https://www.epa.gov/risk-communication).

Over the years, the Intergovernmental Panel on Climate Change (IPCC) has gone from being a typically poor communicator to being a model communicator. Look at their recent reports and slide decks which are attractive and engaging and clearly convey sources, methods, and conclusions. Note that they provide the same information in several different forms to serve different audiences but all hyperlink to the primary assessment so that evidence to support a statement is readily available. Download their visual style guide and let it teach you to better use graphics in reports and presentations (IPCC WGI Technical Support Unit 2018). Ideally, you will have access to a professional editor and a graphic artist. Your ideal editor should be able to go beyond grammar and formatting to help you write clearly for your audience. Your ideal graphics person should be experienced in data visualization and not just computer graphics software. In my experience, such ideal communication professionals are rare, so you probably will need to improve your own skills to work with whatever editorial and graphics support is available.

10

Pollution Control and Risk Management

You might think it is sufficient for protection of the environment to register and permit chemicals and preclude those that pose excessive risks. However, contamination by discharges to water and air, and other waste streams still must be controlled. This is essential because of chemicals that were "grandfathered in" because they were in use before registration was required. It is also required for chemicals that were permitted without sufficient information. In addition, naturally occurring chemicals, such as those in mine waste leachates, are not registered. That is, if an individual hazardous agent is not adequately regulated, then water, air, and soil must be protected from toxic releases.

Diverse strategies are employed for pollution control. Each is implemented in different ways in different countries, and they are subject to many regulations and guidance documents that may not prescribe consistent approaches or methods. In this basic introduction, I describe each approach, its advantages and disadvantages, and the role of assessors in each. I emphasize water pollution because in the U.S. and most other jurisdictions, laws and regulations for water are stronger and the methods more diverse than for other environmental media. Different agencies and different laws and regulations within agencies use various terminologies. I use the term standards here to indicate enforceable benchmarks, but other terms are used in specific jurisdictions.

10.1 Technology-Based Standards

The most straight-forward strategy is to require a type and level of treatment of an emission. The technologies are variously defined as: best practicable control technology currently available, best conventional pollutant control technology, best available technology economically achievable, or other such terms. For example, in the U.S., municipal sanitary sewage must receive secondary treatment which achieves permitted levels of biological oxygen demand, total suspended solids, and pH. More than 50 industries have their own USEPA technology requirements for aqueous effluent permitting. For the criteria air pollutants, the USEPA designates New Source Performance Standards (NSPS) that include equipment and operation specifications. The

DOI: 10.1201/9781003156307-11

primary advantage of this technology approach is that the requirements are known to a prospective emitter and can be achieved at a reasonable cost. The primary disadvantages are that available technologies may not be protective of human health or the environment, and this approach creates no incentive to develop more effective technologies. This approach requires little assessment other than scaling the treatment to the waste and receiving water flows. The technology-based strategy is also applied to solid waste disposal. For example, municipal solid waste dumps in the U.S. must have liners and caps that meet standards.

10.2 Emission Standards

An obvious control strategy is to specify permissible concentrations or release rates of chemicals and other agents in emissions. The responsible party is required to monitor and report results for a list of chemicals and physical properties such as pH, suspended solids, conductivity, and temperature. The permissible levels may be set for end-of-pipe, or they may be adjusted for flows in the receiving waters. For example, emissions may be regulated at the edge of a zone of initial dilution. (This practice is increasingly discouraged in the U.S.) The major advantages of emission standards are that they are straight-forward and that responsibility is clear. Major limitations include the relatively small number of concentrations and properties that typically are reported, the failure to consider mixture effects, and the permitting of conventional pollutants and not those that are specific to the source. For example, at Oak Ridge National Laboratory, in the early years of regulation, we were frequently cited by the state for excessive chlorine emissions but not for the radionuclides emitted. Emission standards also do not address unpermitted and nonpoint sources such as pesticide applications, pavement runoff, erosion, and spilled or buried wastes. Finally, they do not address the combined effects of multiple sources. Assessors are involved in deriving the permissible levels, in writing the permits for new sources, and in reviewing the efficacy of permits.

10.3 Comparison to Ambient Media Standards

Humans and other organisms are exposed to pollutants in ambient water, air, soil, and sediment, not in effluents. Hence, it makes sense to regulate pollutants in ambient media. For U.S. waters, this is referred to as water quality-based toxics control. The USEPA and other agencies have devoted much effort to deriving and applying ambient water and air quality standards to implement this approach. In general, acute and chronic values are applied to protect aquatic life from short-term and long-term exposures (Chapters 4, 21). In addition to concentrations, standards must specify the temporal dimensions—averaging time and recurrence frequency. Assessors are engaged in deriving ambient standards, critiquing them for industries, and applying them to individual cases. Application involves statistically analyzing data from ambient measurements or modeling ambient concentrations, temperatures, pHs, etc. Regulatory agencies provide guidance on sampling and analysis of media for implementation of standards or other benchmarks.

Human health standards are available for ambient water and air. Ambient water criteria for human health are based on consumption of untreated water, edible fish, and invertebrates or recreational contact. They are derived from conventional human health benchmarks, models of bioaccumulation by aquatic biota, and human consumption rates. Drinking water standards for water utilities are more numerous, simpler to derive and apply, and more relevant to most consumers. Benchmark concentrations of some bioaccumulative chemicals in foods such as fish and shellfish are used to protection of humans and relevant wildlife.

Comparison of concentrations in contaminated media to ambient media benchmarks is also the most common approach to contaminated site remediation (Chapter 22). That is, wastes and contaminated soils and sediments are removed, capped, or otherwise remediated if relevant benchmark concentrations are exceeded.

10.4 Identifying Toxic Effluents or Wastes

The toxicity of effluents or wastes may be tested by exposing organisms to the whole material. The test results indicate whether the material should be released, disposed of, treated, or otherwise regulated. In particular, whole effluent toxicity (WET) testing is sometimes used in the U.S. in aqueous effluent permitting. Freshwater WET testing includes conventional acute tests and subchronic (7 day) tests of a planktonic crustacean (*Ceriodaphnia dubia*), the fathead minnow (*Pimephales promelas*), and rarely an alga (*Selenastrum capricornutum*). Other species are used for saltwater effluents. The standard effluent test protocols developed by the USEPA were notable for the degree of field validation, particularly the 7-day subchronic tests (USEPA 2002a, f, g). WET testing may be used to assure that the effluent is not toxic or, with dilution, to assure that the receiving waters are not toxic.

The obvious advantage of these tests is that they address mixture toxicity of the actual effluent, including chemicals and properties for which there are no standards. Like other sets of tests, it may not protect taxa, life stages, and responses that are sensitive to particular contaminants. For example, the fathead minnow 7-day larval survival and growth test is a reasonably sensitive fish test but does not include reproduction, behavior, or other sensitive responses or sensitive species. Mammalian effluent and waste tests were considered in the 1980s, but were judged to be impractical, so human health risks are not addressed.

If an effluent is toxic, the cause of the toxicity must be identified to determine how treatment must be altered. This process is called Toxicity Identification/Evaluation (TIE) (Chapter 11). A variety of chemical techniques are used for this purpose including pH adjustment, dechlorination, aeration, filtration, chelation, fractionation, spiking, and separation to create waters to be tested. The USEPA has provided guidance for TIE and some industries have developed TIE methods for their effluents, but creative chemistry may be required to isolate the causal chemicals in unconventional effluents. TIE is followed by Toxicity Reduction Evaluation (TRE) which is the engineering process of determining how to eliminate an effluent's identified toxicity.

Whole solid wastes such as ashes or mine spoils and their leachates or pore waters may be tested to determine their toxicity. In addition, if contaminated sediments are dredged or contaminated soil is excavated, they are considered wastes and may be tested before disposal is permitted.

Test plants and animals may be exposed to atmospheric emissions diverted into fumigation chambers. However, such tests are not used routinely.

10.5 Ambient Media Toxicity Testing

Testing of ambient waters may be used to regulate water quality for a reach or watershed with single or multiple sources under the USEPA Total Maximum Daily Loads (TMDL) program. The aqueous tests are usually the same as those used for effluent toxicity testing (above). The media may be collected on a gradient away from the source to determine the extent of toxicity and its relationship to contaminant concentrations. In some cases, the tests may be performed in place. For example, plants may be planted on a contaminated site or caged fish may be placed in a water body.

10.6 Biological Standards

Most biological standards in the U.S. are based on aquatic invertebrate or fish community indices or simpler metrics such as species richness. Invertebrate communities often include the most sensitive species, and they occur in waters that are too small to support a fish community. However, there are exceptions such as selenium which is more toxic to fish and herbicides that are more toxic to plants. Biological standards for terrestrial communities have been suggested but have not, to my knowledge, been established in regulatory programs. Biological standards are inherently realistic, but they must be combined with other evidence to determine the cause of any observed impairment (Chapter 11).

10.7 Narrative Standards

When pollutants have no numerical standards and include substances and properties that are difficult to specify or quantify, narrative standards can be applied. The narrative standards are statements that describe the desired water quality goal, such as waters being free from oil, color, odor, and any substances that can harm people or aquatic life. They serve as catchall criteria that are applied when numerical criteria fail.

Narrative standards may seem weak because they are vague relative to numerical standards, but they can be effective. During the Superfund remedial investigation of the Oak Ridge Reservation, while our attention was focused on mercury, PCBs, and radionuclides, a State of Tennessee employee noticed an oil sheen in a stream. That violation of the narrative standard forbidding oil sheens resulted in the diversion of parking lot runoff to the water treatment plant. In addition to the toxicity of oil leaking from automobiles, pavement runoff can be toxic due to multiple pollutants and heat. That oil sheen incident resulted in significant water quality improvements as confirmed by subsequent biological monitoring.

Another example is provided by the court case in which West Virginia coal companies were sued for allowing leachate from coal mining to greatly increase the specific conductivity in streams (Chambers 2014). The judge ruled that the USEPA benchmark value for conductivity was not a criterion, but it was strongly supported by the scientific evidence. Therefore, where it was greatly exceeded in West Virginia waters, that exceedance was de facto evidence that the State's narrative standards were violated.

10.8 Multiple Standards

Different approaches provide different information. Each approach to pollution control is considered to be sufficient, but they can complement each other and provide more reliable protection. They each rely on certain critical assumptions that do not occur in other approaches (Table 10.1). Different strategies are employed for using multiple standards in a complementary way.

Independent applicability—The simplest way to combine different approaches is to assure that the receiving community is protected by each. For example, when renewing a permit for an aqueous effluent, if the treatment does not meet technology standards, if the effluent or ambient water exceeds a single chemical standard, if the effluent or ambient water is toxic, or if the biotic community is impaired, the effluent requires additional treatment. In other words, if the receiving waters are physically, chemically, or biologically impaired, the emissions are unacceptable. This approach is based on precaution and the premise that false positives are less likely than false negatives.

Validation—Weaknesses in a single approach may be treated by validating their results with an approach that lacks those weaknesses. For

TABLE 10.1

Summary of Approaches to Providing the Information Needed for Pollution Control

Approach	Nominal Result/Unknowns and Problems
Technology based	The technology for your facility will be deemed adequate. No testing and little analysis is required, and common problems are solved at reasonable cost. / Must assume the technology is appropriate in each case.
Emission standards	The effluent is not toxic with respect to the analyzed chemicals. / Must assume that all potentially toxic chemicals are analyzed, there are no combined effects, and the benchmark values are protective of the receiving community.
Ambient media standards	The water is not toxic with respect to the analyzed chemicals. / Must assume that all potentially toxic chemicals are analyzed, there are no combined effects, and the benchmark values are protective of the receiving community.
Effluent toxicity testing	The effluent is not toxic with respect to the test organisms and endpoints. / Must assume that the tests are protective of the receiving community. Should apply TIE to determine the cause of any toxicity.
Ambient media toxicity testing	The water is not toxic with respect to the test organisms and endpoints. / Must assume that the tests are protective of the receiving community.
Biological standards	The water is not adversely affecting the biotic community. / Must assume that the biological criteria are protective of the full community and that other agents (i.e., confounders) are not the true cause.
Narrative standards	Narrative standards fill gaps in numerical standards; goals specified in narratives are achieved. / Must assume that staff will use adequate methods and interpretations.
Multiple standards	Multiple approaches give reliable results. / Time and effort to implement.

The / separates expected advantages from disadvantages.

example, effluent toxicity tests are often used to confirm that the numerical chemical-specific standards are in fact preventing toxic effluents. This is a validation because toxicity tests address the combined toxic effects and the effects of non-criterion chemicals that the chemical standards miss.

Weight of Evidence—Each of the approaches provides different information that may serve as evidence for or against the hypothesis that an effluent or waste is acceptable. The result depends on the sign and weight of the body of evidence and the implications of the available evidence (Chapters 24, 25). Implications of the evidence are:

- Technology standards determine whether the treatment technology is state-of-practice.
- Single-chemicals standards determine whether chemicals are present in toxic amounts.
- Toxicity tests determine whether the effluent or ambient medium is toxic.
- Biological standards determine whether there is a biological effect in the receiving system.

A weighing of the available evidence can indicate whether the hypothesis is supported and consideration of the implications of the evidence can explain any apparent contradictions in the evidence.

10.9 Accident Prevention

Accidental releases of hazardous materials can be minimized by regulations that prescribe protective technologies and actions. The technologies include double-hulled tankers, strong storage tanks that can collect leaks, and blow-out preventers on oil wells. Actions include safety training, speed limits, evacuation plans, and regular inspections of equipment. Such regulations are developed and enforced by agencies like the U.S. Department of Transportation. The USEPA was ordered by a federal court to develop rules to prevent and contain spills of oil and hazardous substances as required by the Clean Water Act. In 2019, the Trump administration released a regulation stating that the EPA would be "not establishing at this time new discharge prevention and containment regulatory requirements under CWA section 311," stating that it was unnecessary (USEPA 2019a). Regulations to minimize spills are derived and implemented on the basis of assessments of the frequencies and severities of different types of accidents and of the efficacy of technologies in reducing risks.

10.10 Risk Management

Risk management is the process of deciding what action should be taken given the results of a predictive risk assessment and non-scientific considerations. The strategies discussed above are the links between the biology and chemistry of health and ecological assessments and the decision of what the management goals and processes should be. However, the final risk management decision is also informed by several factors (USEPA 2022e):

- Economic factors, such as the benefits of reducing risks, the costs of mitigation or remediation options, and the distribution of costs and benefits.
- Laws, regulations, and legal decisions.
- Social factors for human health risks such as income, ethnicity, values, and land use.
- Technological factors that limit the range and feasibility of management options, as well as risks from each technology.
- Political factors including the input from local, state, and national governments.
- Current policies and practices, as defined by any responsible agency and input from stakeholder organizations.
- Public values and attitudes about the environmental risks and risk management options.

These factors are considered by risk managers in consultation with assessors, economists, and decision support staff.

Part II

Concepts of Environmental Assessment

The following chapters explain ten basic concepts that are fundamental to environmental assessment. Some, such as causation, are big essential concepts that philosophers and scientists struggle with. Some such as probabilistic assessment are contentious but not as important as they seem. Others such as integrated assessment are not controversial but are difficult to implement. All are worth a chapter because they can trip up assessors.

DOI: 10.1201/9781003156307-12

11

Causation

I would rather discover one causal relationship than be king of Persia.
—*Democritus (430-380 BCE)*

You may think that the concept of causation is obvious, but it may be the most important and difficult concept in philosophy and science. We are taught that correlation is not causation, but we are not taught what is causation. This is important because every environmental assessment has a causal question at its center. What caused this observed effect? What effects does or will this chemical cause? What concentration of this chemical will cause no significant effects?

Causal assessments (Chapter 1) are at least as complex as risk assessments but are seldom recognized as distinct. This chapter presents only some basic principles of causal assessment and a few important approaches. I encourage you to obtain the open access book Ecological Causal Assessment (just Google the title).

11.1 Types of Causation

The history of causation is littered with different ideas of how we know a cause when we see it and even whether causes exist at all. Three fundamental concepts that are still in use are the associationist, interventionist, and counterfactual theories.

11.1.1 Associationist Causation

David Hume, the Scottish enlightenment philosopher, was the first to seriously analyze how we determine causation. "We have no other notion of cause and effect, but that of certain objects, which have always conjoin'd together, and which in all past instances have been found inseparable" (Hume 1739). He argued that association proves nothing, but it is all we have. Statistics retains the associationist theory of causation but allows for inconsistent association. Karl Pearson (1896), a pioneer of statistics, stated that correlation

DOI: 10.1201/9781003156307-13

captures all knowledge of causation. Most statisticians since then have continued to develop methods of quantifying associations, without concern for further demonstrating causation. For example, regression is linear correlation, so it demonstrates only association. Bayesian networks add the directionality of causation but are still just probabilistic associations (Pearl and Mackenzie 2018).

Associations are not causal for two reasons. First, except for causal networks, they are not directed. If two variables are correlated either one may be the cause of the other. Second, the correlation may be due to a third variable that is confounding the relationship (Section 11.2.2). Correlation is evidence of some causation; you just cannot be sure what is causing what.

The effects and apparent causes in observational studies used in epidemiological and ecoepidemiological assessments are, unlike experimental results, real-world phenomena. Although many assessments stop after determining association of effects with a hypothesized cause, causal assessments should continue by considering the direction of causation and the influence of particular confounders. For example, a synoptic study of stream invertebrates in West Virginia found that reduced mayfly diversities were associated with valley fills and high specific conductivity (Pond et al. 2008). A separate causal assessment was required to determine that specific conductivity was causal and not significantly confounded (Cormier et al. 2013; Suter and Cormier 2013).

11.1.2 Interventionist Causation

The only reliable method for determining causation is a well-designed and well-conducted experiment. That is, we intervene in controlled systems by introducing an agent and, because we have control of that potential cause and have randomly assigned it to receptor systems, we know that it caused the observed differential changes. If there are no critical errors in design or performance, control of exposure and randomization will ensure that the effects are caused by the exposure plus some random variance. Clinical trials of drugs represent this ideal—the subjects are people, the conditions are medical treatments of a defined disease, and the exposures are randomized and blinded. However, we cannot dose people with environmental contaminants. We expose rodents to represent people and expose various test species or model ecosystems to represent ecological endpoint entities. As a result, the cause–effect relationship in the experiment is not exactly the one we want. We must extrapolate from the laboratory to the field, species to species, one exposure regime to others, and one level of organization to other levels (Chapter 26). Also, exposure forms and levels in tests may not be the same as the exposures in the field. But at least the experimental results are caused by the treatment. However, this method alone is not very reliable determinate of causation in the field due to the extrapolations listed above.

Evidence of interventions may also be important in observational studies. If you are called to investigate a fish kill, you may be at a loss if you start looking for associations with the many potential causes. You may focus by asking what has been done to the stream that was not done to other streams in the area? If you know of an intervention such as a pesticide application in an adjoining field, that preceded the deaths and an absence of deaths in streams without adjoining pesticide applications. That evidence can focus you on pesticide residues, characteristic symptoms, and sensitive species. The pesticide application becomes, in effect, an unplanned and unreplicated treatment (aka a natural experiment).

11.1.3 Counterfactual Causation

Counterfactuals are statements of the form, if C had not occurred, E would not have occurred. Therefore, C was the cause of E. We cannot empirically verify counterfactual causation because, by definition, the cause did not happen. It is based on knowledge of the mechanisms that define the system of interest. We know that no matter how many times a rooster crowing and sunrise are associated, if the rooster had not crowed, the sun would still have risen because there is no potential mechanism by which a rooster could influence sunrise. Although you cannot test or observe counterfactual causation, you can model it if you have sufficient understanding (Section 11.2.4).

Counterfactual arguments are subject to preemption and overdetermination. For example, an accident at a factory may release lethally hot water due to failure of the cooling system. The same accident may bring down the wastewater treatment plant so that acutely lethal untreated effluent was released. We cannot say that the either release was a counterfactual cause of a fish kill because both releases were capable of killing aquatic organisms. This is an example of overdetermination. Preemption is similar except that one potential cause occurs before the other and preempts it.

11.2 Methods for Causal Assessment

Many methods have been used to assess causation, and many of them do not actually determine causes. The following five methods will work if you have the necessary information.

11.2.1 Weight of Evidence

Weighing all available evidence for a causal hypothesis has been generally recognized as the best available approach to causal inference ever since Hill published his causal considerations (Section 24.5). Weight of evidence can use whatever qualitative or quantitative evidence is available. It can even mix associations with interventions and counterfactuals.

For several reasons, Hill's considerations are inadequate and are being replaced by more defensible methods such as the USEPA method in Chapter 25. The part of the USEPA method that is particular to environmental causal assessment is a list of potential implications of the evidence called characteristic of causation (Cormier, Suter, and Norton 2010; Cormier, Suter, and Norton 2014; USEPA 2016e). Table 11.1 is a weight-of-evidence table that uses those characteristics to summarize the evidence for major ions as a cause of biological impairment of Central Appalachian streams.

Co-occurrence—The cause co-occurs with the entity in space and time. Ideally, continuous monitoring has demonstrated that the causal agent is present when the effect occurs. However, in most cases, you must use measurements taken after the effects are observed and assume, based on evidence, that current co-occurrence implies co-occurrence since the induction of effects. In some cases, to identify co-occurrence, you must allow for movement of the causal agent or affected organisms or for time lags in the induction of effects.

Sufficiency—The intensity, frequency, and duration of the cause are adequate for the affected entity to exhibit the observed type and magnitude of effects. This evidence comes primarily from toxicity tests for chemicals or equivalent studies of other agents. However, evidence for sufficiency of some agents such as siltation of streams comes from field studies because they are very difficult or impossible to replicate in a laboratory.

TABLE 11.1

A Summary of the Weight of Evidence for Each Causal Characteristic in Support of the Hypothesis That Major Ion Activity, Measured as Specific Conductivity, Is the Cause of Observed Loss of Invertebrate Genera in Central Appalachian Streams

Characteristic	Evidence	Score
Co-occurrence	The loss of genera occurs where conductivity is high even when potential confounding causes are low but not when conductivity is low.	+++
Preceding Causation	Sources of conductivity are present and are shown to increase stream conductivity in the region.	+++
Interaction	Aquatic organisms are directly exposed to dissolved salts. Physiological studies document effects of ion imbalance.	+
Alteration	Characteristic genera are affected at sites with higher conductivity.	++
Sufficiency	Specific conductivity and duration of ion exposures affect invertebrates similarly in both field and laboratory analyses.	+++
Time order	Conductivity increases, and local extirpation occurs after mining permits are issued, but before and after data are not available.	NE

The number of+symbols represents the weight of evidence.

Time order—The cause precedes the effect. This is the only absolute causal characteristic. Effects never precede their causes. However, we often do not know when the cause began relative to the effect.

Interaction—The cause interacts with the entity in a way that can induce the effect. The most common types of evidence of interaction with chemicals are body burdens or biomarkers of exposure. Other agents have their own evidence. For example, gill abrasion is evidence of interaction with suspended sediment.

Alteration—The entity is changed by the interactions with the cause. These alterations may be characteristic symptoms such as convulsions, cranial deformities, or gasping at the surface. They may also be higher level changes such as loss of species known to be sensitive to a potential cause.

Antecedence—The proximate causes are results of a series or web of prior causal events or agents. In particular, a hypothetical association of cause and effect is more credible if there is evidence of a source or a route of transport. In some cases, finding a source is sufficient to identify the cause (Bellucci, Hoffman, and Cormier 2010). Also, knowledge of antecedence may identify confounding factors that must be controlled in any quantification of causal strength.

You need not have evidence for all of these characteristics to reach a conclusion of causation. Still, in general, a body of evidence has more weight if it supports more characteristics. In addition, the characteristics can serve as a guide to the types of evidence that you might find or generate to make your case.

A distinct type of environmental weight of evidence is pathology focused. If a kill of wildlife or fish is found, the carcasses may be passed to a veterinary pathologist who will perform autopsies to check for injuries, symptoms of disease or poisoning, pathogens or chemicals in tissues, and other diagnostic signs. In many cases, the pathologist will weigh this body of evidence and identify the cause. However, success is more likely if pathology is just one part of a wider view of evidence and evidence weighing. A mixture of pathology and more ecological evidence was essential in the San Joaquin kit fox causal assessment (Suter and O'Farrell 2009). The weighing of evidence by a pathologist is common in human deaths, but the sort of acute effects that usually prompt an autopsy are relatively rare in environmental assessments for human health.

11.2.2 Causal Network Analysis

I believe that weighing the body of evidence is the best general approach to assessing causation, but causal network analysis is becoming an important quantitative technique. In particular, it can quantify the strength of causal relationships in the network. (The use of these networks for modeling in general is discussed in Chapter 23.) Causal network analysis began with Sewall Wright's path analysis, but Judea Pearl has applied Bayesian statistics and popularized it as a representation of causation. Pearl made us aware that analysis of causation should be constrained by our knowledge of the structure of causal relationships in a system (Pearl and Mackenzie 2018). In particular, when we diagram causal relationships in a network model (Chapter 23), the arrows between two boxes go only one way and there are no cycles because that is the way causation works. The diagrams are called directed acyclic graphs (DAGs). Conventional statistics is bidirectional (A is correlated with B to the same degree as B is correlated to A), so it is not causal. Pearl created Bayesian network modeling to solve that problem, but he later realized that he had not solved the problem of confounding. Therefore, Bayesian conditional probabilities linking A and B were not necessarily causal and could, in fact, be caused by C (Figure 23.2). In the real world, you may have many variables directly or indirectly confounding your estimate of the strength of the cause of interest. The usual solution to confounding in observational studies is to condition on each variable to determine whether it significantly changes the result. For example, in epidemiology, subjects are typically stratified by age class, sex, and income. However, depending on the relationships of the other variables with the cause of interest, those results may be misleading. Pearl provides rules for identifying the proper set of adjusting variables (Pearl and Mackenzie 2018).

Another successor of Wright's path analysis, structural equation modeling, provides a frequentist analytical approach that is more familiar than Pearl's Bayesian networks. Unlike conventional regression analysis, the structural equations represent the structure of a causal network (Shipley 2000). Pearl's tricks with networks apply to these analyses as well, and for linear relationships, the adjustments are simple. In terms of causal status, Bayesian networks and structural equation models are equivalent.

The basic problem in developing causal network models is ensuring that they in fact represent the causal processes in the system being assessed. The network diagram is the starting point of quantitative causal network modeling, but Pearl and other network modelers have little or nothing to say about how to develop the diagrams. Their advice is to find experts. You should take that suggestion seriously but also recognize that there is a method for deriving causal networks, weighing the evidence for the causal links (Chapter 24).

The point of this section is to make you aware that, except for being directional, Bayesian networks are not inherently causal. Both Bayesian networks and structural equation analyses require learning new terminology and techniques (e.g., Pearl's back door criterion) to ensure that the results are causal.

11.2.3 Toxicity Testing for Causation

Toxicity Identification Evaluation (TIE) is a useful but specialized method for determining causes of toxic effects (Norberg-King et al. 2005; USEPA 1999). If you have determined that an effluent or ambient water is toxic, you need to know what constituents cause the toxic effects. Otherwise, you do not know what treatments to implement or what sources to regulate. You do that by fractionating the mixture and testing the fractions, adding components of the mixture to the background medium and testing each component, adjusting the water chemistry (e.g., pH and dissolved oxygen), or other techniques. TIE was developed for regulation of effluents (i.e., what constituent of an effluent should be regulated?—Chapter 10), but it can be used more broadly to answer causal questions about constituents of mixtures. For example, the cause of acute mortality of coho salmon migrating into Seattle, Washington, urban streams was determined to be an antioxidant chemical in tire wear particles by a study employing TIE and some fancy analytical chemistry (Tian et al. 2021).

11.2.4 Counterfactual Modeling

A common concern of the public and policymakers is to identify the cause of extreme weather events. Was that hurricane, drought, heat wave, or flood caused by climate change or not? Several groups are devoted to answering these questions such as the World Weather Attribution collaboration (Philip et al. 2020). One approach is to empirically model changes in the frequency of events of various intensities in a certain region before and after significant increases in greenhouse gases. That approach requires sufficiently frequent events, a long weather record, and an assumption that there are no confounders such as volcanic eruptions. A more focused approach is to mechanistically model the thermodynamics of the climate system that caused the event, and run it with and without the cumulative heat input since 1900 or some other baseline. That approach requires more time, effort, and expertise. These are both counterfactual approaches; the counterfactual being frequencies of extreme events in the region without global warming.

You may address less dramatic causal problems by counterfactual modeling. For example, would an observed decline in a riverine shad fishery have occurred in the absence of entrainment in power plant cooling systems, harvesting, or pesticide exposures? You can answer these questions by modeling the fishery without the power plant, the fishery, or pesticide toxicity.

11.2.5 Outcome Assessment and Causation

Outcome assessments are conducted primarily to assure that environmental contamination or impairments have been eliminated by remedial actions or other interventions (Chapter 1). However, they also provide the final evidence concerning causation. If recovery is possible (e.g., there is physical habitat and a source of potential colonists), and you reduced the contaminants of concern to nontoxic levels, you expect the biotic community to recover. If recovery occurs, that result of the intervention supports your conclusion that the contaminant was the cause. If not, it is evidence that a different cause or an additional cause was responsible. That should prompt you to perform a new causal assessment.

A causal outcome assessment is not as simple as that introduction suggests. You seldom do just one thing when managing the environment and any change may be a cause. Altered treatment of an effluent changes its properties besides reducing the concentration of the contaminant of concern. Remediating contaminated media may not restore the biotic community because of physical damage such as removal of soil or sediment or the use of dispersants. Even if the remediation is perfect, lack of dispersal of organisms to the site may prevent recovery. Hence, careful attention to alternative causes is important.

12

Hazard

Hazard is the potential of an agent to cause a particular harm when receptor organisms or systems are exposed to it. When assessing risks from exposure to a chemical or other agent, you must determine what hazard or hazards you will assess given the case. This process is called, not surprisingly, hazard identification. It may be completed during the problem formulation in simple cases, particularly those cases that are well specified by the charge to the assessors (Chapter 4). Hazardous properties of chemicals may be found in databases (Chapter 21). For example, if the chemical of concern is identified as a human carcinogen in an IRIS assessment, your problem formulation may accept that designation. However, if the hazard is unknown or ambiguous, the hazard identification requires analysis and weighing of evidence, which is usually included as the first step in the analysis of effects (Chapter 7).

12.1 Hazard Identification and Assessment

Hazard identification is part of a risk assessment, but a hazard assessment is a distinct document that provides a generic review of the hazardous properties of a chemical or other agent. When available, they may be used as a source of information for your assessments. The OECD (2022b) provides consensus hazard assessments for industrial chemicals. Hazard assessments may also derive benchmark values (Chapter 22).

A hazard identification defines what agents may cause what effects, on what endpoint receptors, by what routes. Hazard identification is primarily devoted to identifying the potential effects, but you need more information to derive the exposure–response relationship.

- Hazardous agents are often defined by the charge for the assessment, but in some cases, such as chemical mixtures, waste sites, or effluents, specifically defining the agents of concern is part of the assessment.
- Hazards happen to a receptor. You may begin with the receptors identified as assessment endpoints because of their importance (Chapter 14). However, information concerning the hazardous agent is required when considering the susceptibility of potential endpoint

DOI: 10.1201/9781003156307-14

receptors. Therefore, assessment endpoints may be added or refined following hazard identification. The endpoint receptor is usually not an issue in human health assessments because it is individual humans within an exposed population. However, in some cases, such as the risk of preeclampsia from boron in pregnant women, categories of human receptors are specifically defined.

- Effects are changes in attributes of endpoint receptors. They include the attributes of assessment endpoints, but may also include mechanisms. For example, death by cholinesterase inhibition is more informative than simply death.
- Different routes of exposure may cause different effects. For humans, the routes are ingestion, inhalation, and dermal. For ecological receptors, the potential routes are direct aqueous and sediment contact (aquatic and sediment organisms), direct soil contact (soil invertebrates and terrestrial plants), soil or sediment ingestion (some terrestrial animals and sediment and soil invertebrates), dietary ingestion (any animal), and direct atmospheric inspiration or uptake (mainly terrestrial plants but also animals).

In human health assessments, the key hazard distinction is between carcinogens and noncarcinogens. Hazard identification for declaring that a chemical is a carcinogen may require major efforts, and their results are often contentious. For example, almost 50 years after DDT was designated a possible human carcinogen, that hazard is still controversial. Ecological assessors have distinguished acute lethality from chronic effects (Box 7.1) and have often left the effects in hazard identification at that. Recently, both human health and ecological assessors have paid more attention to distinguishing more specific effects, particularly reproductive and developmental effects.

The same conventional laboratory test data and observational field data used in risk assessments are the most common input to hazard identification. The difference for hazard identification is that there need not be sufficient data to define an exposure–response relationship. The study results need only be qualitative, not quantitative, and need not be distributed over a range of effects levels. Also, if it is clear that the agent of concern is a cause of a particular effect in the field, for hazard identification you need not worry that a co-occurring agent is also contributing to the effect.

Because hazard identification does not require a quantitative association of the agent with an endpoint effect, mechanistic information is more useful than it is for risk assessment. These may include symptoms displayed by test organisms or physiological, histological, and molecular data. Even in vitro studies may be sufficient to identify a hazard if an Adverse Outcome Pathway (AOP—Chapter 23) is available. AOPs are conceptual models of the processes connecting a molecular initiating event to an adverse outcome. Their great advantage is that in vitro tests can determine that a chemical causes a particular molecular initiating event, such as binding to a hormone receptor, and the AOP then predicts the adverse outcome. So, in principle, if you know whether a chemical causes a particular initiating event, an AOP allows you to determine that chemical's hazard. The adverse outcomes are typically organism level effects, but extension to population-level effects has been demonstrated.

12.2 Hazard Matrix

Impact matrices were developed for impact assessment to address the many agents and activities potentially caused by a project and the many environmental entities and processes that could be exposed. The matrices have rows for project activities and columns for environmental receptors. The U.S. Geological Survey's standard Leopold matrix had 100 rows and 88 columns (Leopold et al. 1971). For hazard identification, the cells of the matrix could be simply checked if they were judged to represent a hazard. However, the Leopold matrix includes numerical scores for magnitude and importance of potential impacts. In some impact assessments, those scores are the sole analysis.

The matrix approach could be used for hazard identification in complex environmental risk assessments. Checking matrix cells would be sufficient in many cases, but weights for the potential interaction could prioritize the hazards.

13

Integration of Human Health and Ecological Assessments

One of the first conditions of happiness is that the link between man and nature shall not be broken.

—*Leo Tolstoy (1847–1910)*

Most institutions that perform environmental assessments organizationally separate those who address ecological effects from those who address human health effects. This is referred to as stove piping. Although it is traditional to treat humans as separate from the environment, in environmental assessment, it is advantageous to integrate them.

13.1 Advantages of Integration

Consistent results—An important motive for integration is avoidance of inconsistent results. For example, in the assessment of East Fork Poplar Creek, in Oak Ridge Tennessee, the ecological risk assessment found that mercury was not taken up by plants on the flood plain, but the human health assessment used default inorganic mercury uptake values to estimate human exposure. Hence, the ecological assessment concluded that herbivores were not significantly exposed, but human health assessors concluded that hypothetical future gardeners would be significantly exposed.

Common concepts—Assessors who share basic concepts will collaborate more effectively and will benefit from sharing methods and approaches. For example, we ecological assessors created the term ecoepidemiology as an analogy to epidemiology to describe the characterization of ecological effects in the field and the determination of their causes. We also borrowed epidemiological concepts such as Hill's considerations.

Common narratives—Human health and ecological assessors should achieve a common understanding of their results so that they provide a common description of what is happening or will happen in a case. Even when all assessors use the same information and assumptions, the interpretation of results may be different and that creates problems. At least, be prepared to

DOI: 10.1201/9781003156307-15

explain to a landowner how a stream that is ecologically impaired is not a health threat to his children who play in it.

Efficiency from collaborative activities—Some measurements, analyses, and modeling activities are common to human health and ecological assessment and need not be duplicated. This requires collaboration and agreement on issues such as soil sampling depths, filtration of water samples, and the accident scenarios to be modeled.

Interdependence—Some effects on humans result from effects on the nonhuman environment. For example, contamination of fish and shellfish can result in human dietary exposure, loss of livelihood, or loss of nutrition. Deformities of oysters exposed to tributyl tin made them unmarketable. The Yupik people of the Bristol Bay Alaska watershed depend for both subsistence and income on the salmon which were threatened by the proposed Pebble Mine. The literature on services of nature addresses this interdependence from the ecological perspective (Chapter 14).

Best information—By sharing information, environmental assessors can use the best available information, even if it comes from the other side. In general, ecological assessors have more information on what is actually occurring in the environment, and human health assessors have more information on suborganismal mechanisms of the agents of concern. The most prominent example is DDT. Ecologists determined that it persisted and biomagnified in food webs leading to studies of differential exposure of human populations. That led in turn to concern for exposure to persistent and bioaccumulative chemicals in general.

Optimum decision—A contaminated site can be remediated one way. There is no human health remediation and ecological remediation unless there are alternative universes. If the biotic community on a site is destroyed to protect human health because of an unreasonable exposure scenario (e.g., what we called the stupid homesteader), demonstrably incorrect environmental chemistry, or highly conservative human toxicity values, that may not be the optimal remediation. The loss of a mature floodplain forest with a diverse flora and fauna that is showing no effects of the contamination is a loss for humans as well as for the environment. Integrated assessment can, at least, clearly explain the consequences of remedial actions.

13.2 How to Integrate

Use common processes—For decades, USEPA assessors used different assessment processes for human health and ecology. However, with the publication of the 2014 framework for human health risk assessment, the assessment processes and much of the terminology converged. This opened opportunities for integration. In particular, human health assessors now follow ecological

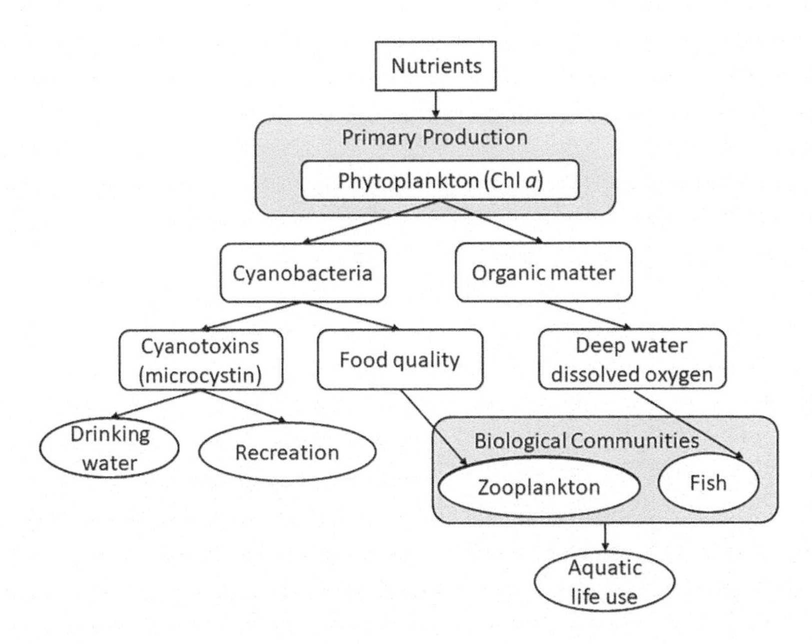

FIGURE 13.1
A high-level network model for the development of ambient water quality criteria for nitrogen and phosphorus. It includes causal linkages of aquatic life (zooplankton and fish) and human welfare (drinking water, recreation, and aquatic life uses) endpoints. (Redrawn from USEPA 2021a.)

assessment practices, particularly by performing problem formulations. That critical step can be performed collaboratively which should increase the consistency of results. In particular, development of common conceptual models can increase consistency and efficiency and reveal causal linkages (Figures 13.1 and 4.1). In the extreme, a single exposure model linked to a common system for modeling effects could be employed (Hines, Conolly, and Jarabek 2019).

Share data, methods, and results—This admonition should go without saying, because human health and ecological assessors should share the same mission. However, there is a tendency to stay in one's silo. Get to know everyone working on an assessment and take the time to share progress and findings.

Consider sentinel species—Nonhuman organisms may act as sentinels for human exposure and effects.

- They are, in general, more exposed to environmental contaminants because they are immersed in the environment and they are more likely to obtain most or all of their food and water from a contaminated site.
- Some of the many nonhuman species are likely to be more sensitive than humans.

- They do not have occupational or lifestyle exposures that confound causation.
- They can be readily sampled, analyzed, and necropsied.

A classic example is the appearance of severe mercury poisoning of animals in both Minamata, Japan, and the Everglades, Florida, before there was awareness of human exposure or toxicity. Effects on nonhuman organisms in the environment should serve as an alert, just as laboratory rats serve as models for humans. However, routes of exposure and mechanisms of effects must be relevant. Humans do not ingest lead shot or drink from waste ponds, and some effects are not directly relevant to humans. For example, the diclofenac that kills vultures apparently does not harm humans at realistic exposures.

Link humans to the environment—In conventional human health assessment, the environment is just a source of chemicals and other agents that might kill or injure you. However, the environment also provides everything from clean water and food to spiritual experiences. That is, it is essential to human welfare and health is one component of welfare (Figure 13.2). Effects on human welfare are assessed under the U.S. Clean Air Act because it is legally required, but it is done by ecological assessors. Ecological assessors developed the concept of ecosystem services to demonstrate the relevance of ecological effects (Chapter 14), but someone must take the next step to assessing human welfare. Economists address part of it but neglect most of the welfare benefits of the environment. In the USEPA, economists are few and they largely focus on the benefits of avoiding premature human deaths. When appropriate and possible given resource limitations, broader social science

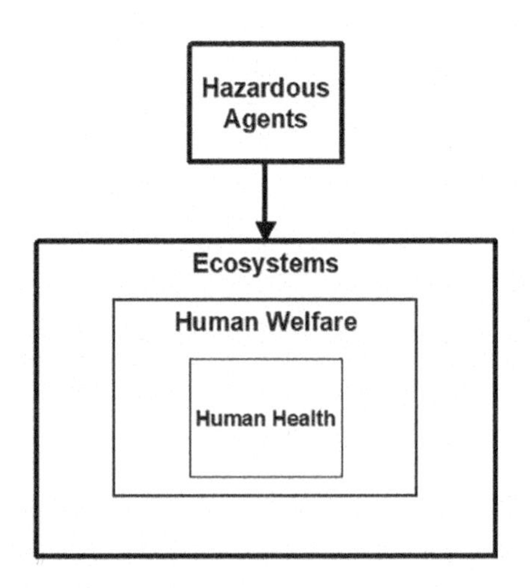

FIGURE 13.2
A diagram illustrating how contaminant sources release contaminants into ecosystems which are affected. Ecological effects reduce human welfare and those welfare effects plus direct human exposure in the altered environment affect human health (Suter 2004).

expertise should be engaged. For the Bristol Bay assessment, we engaged cultural anthropologists to assess the potential effects on the Yupik people of loss of salmon. We also engaged resource economists to assess the monetary value of the salmon to all cultures. That degree of integration is unusual in environmental assessment (EC 2003a).

13.3 Current and Future Integration Practices

The World Health Organization's International Program on Chemical Safety advocated integrated assessment, proposed a framework, and presented case studies (WHO 2001; Suter et al. 2003). The European environment and health strategy calls for an integrated approach to research, information, understanding of pollution, intervention, and stakeholders (EC 2003b). However, the institutional stovepipes are still largely in place. It is up to assessors to collaborate and assure the benefits of integration. At least, when briefing management or stakeholders, you do not want to contradict each other.

Collaboration and sharing to achieve consistency and concordance should be easily achieved, and in some instances, it is. However, working at the interface between humans and the environment is difficult. Institutional processes are needed to bridge the gap between ecologists and social scientists.

14

Goals and Endpoints

The Congress recognizes that each person should enjoy a healthful environment and that each person has a responsibility to contribute to the preservation and enhancement of the environment.

— *U.S. National Environmental Policy Act of 1969*

14.1 Goals

Goals tell you what should be accomplished by the decision that your assessment will support. For most environmental assessments, goals are provided by mandates in laws or regulations. For example, the EU's biodiversity strategy aims to "stop biodiversity loss" (EC 2011). That is a relatively clear goal. Most other goals are more ambiguous. For example, the U.S. Clean Water Act's goal, "restore and maintain the chemical, physical, and biological integrity of the Nation's waters," and the European Water Framework Directive's goal of "good ecological status" are open to various interpretations. Some goals are both ambiguous and extremely broad—the goal of impact assessment in Canada is to "foster sustainability."

Operational interpretations of goals are provided by policy and precedent. For example, a policy stated in the Clean Water Act related to the overall goal mentioned above is prohibition of "the discharge of toxic chemicals in toxic amounts." Policies and precedents are, in general, provided by agencies applying the laws and by courts judging those applications. Assessors should understand them as they are applied in their context. When goals and policies require interpretation for a particular assessment, the decision-maker should be consulted.

The existence of legal goals serves an important function for assessors. We scientists have no standing to apply our values, but assessments require value judgments. The goals of environmental laws and regulations provide the values that are served by assessments. We address protection, prevention, preservation, and restoration (value-laden verbs), because laws require it, not because we are pro-environment. Hence, those who say that assessments must be values-free are naïve. Assessments are performed to serve

DOI: 10.1201/9781003156307-16

public values as expressed by elected officials and interpreted by appointed officials and judges. Assessors swim in a sea of values.

14.2 Ecological Assessment Endpoints

An assessment endpoint is an entity and an attribute of that entity that can represent an environmental goal. You should consider the following criteria when selecting assessment endpoints for ecological assessments (USEPA 1998; Suter 1989):

Policy Goals and Societal Values—Because the risks to the assessment endpoints are the basis for decision making, the choice of endpoint should reflect the policy goals and societal values that a risk manager is expected to protect. This is the dominant consideration.

Ecological relevance—Entities and attributes that are significant determinants of the attributes of the system of which they are a part are more worthy of consideration than those that could be added or removed without significant system-level consequences. Examples include the abundance of a keystone predator species, which is relevant to community composition or the primary production of a plant assemblage which is relevant to herbivore abundance and many other ecosystem attributes.

Susceptibility—Entities that are potentially highly exposed and sensitive to the exposure should be preferred, and those that are not exposed or do not respond to the contaminant should be avoided.

Operationally definable—An operational definition is one that clearly specifies what must be measured and modeled in the assessment. Without an operational definition of the assessment endpoints, the results of the assessment would be too vague to be balanced against costs of regulatory action or against countervailing risks. Ecosystem health is not a good assessment endpoint (Suter 1993a).

Appropriate scale—Ecological assessment endpoints should have a scale appropriate to the site or action being assessed. This criterion is related to susceptibility because populations with large ranges relative to the site of concern have low chronic exposures. In addition, the contamination or responses of organisms that are wide-ranging relative to the scale of an assessment may be due to sources or causes not relevant to the assessment. However acute hazards such as waste pits may be small relative to range but still pose a significant risk, particularly if they are attractive (Chapter 6).

Practicality—Some potential assessment endpoints are impractical because good techniques or information are not available to estimate them. For example, there are few toxicity data available to assess effects of contaminants on lizards, standard toxicity tests for reptile are unavailable, and lizards may be difficult to quantitatively survey. Therefore, lizards may have a lower priority than other, better-known taxa. Practicality should be considered only after the other criteria are evaluated. If, for example, lizards are included because of evidence of particular sensitivity, policy goals, or societal values (e.g., presence of an endangered lizard species), then some means should be found to deal with the practical difficulties.

Without relevance to management goals, an assessment will not serve its purpose. Because most environmental decision-makers, at least at first, know little about ecology, they may ask assessors to justify their choice of endpoints. I have been asked more than once to justify the use of aquatic insects as endpoint entities. Their importance can be explained in terms of ecosystem services (see below), but what the decision-maker wants to know is, will I be sticking my neck out if I make a decision based on insects? The USEPA (2016c) has addressed this issue by publishing generic ecological assessment endpoints which are certified relevant. They are organized in terms of the conventional levels of organization (organism, population, and community/ecosystem) and are justified by identified precedents, policy support, and practicality (Table 14.1). Because these endpoints are generic, when applied to an assessment, they must be made specific for the particular location, hazards, and legal contexts. For example, the generic endpoint, kills of organisms, may refer to recurrent episodic deaths of ten or more fish due to pesticide spray drift or to a single event of death of hundreds of seabirds due to an oil spill.

Environmental entities are susceptible if they are expected to be exposed and are sensitive to the effects of the agent. Determining that a potential endpoint is susceptible requires a screening assessment. However, susceptibility in endpoint selection is a matter of plausibility rather than proof and may be subordinate to relevance. For example, in the Bristol Bay assessment, salmon populations were, by consensus, the primary endpoint and were assumed to be potentially susceptible to anything that could physically or chemically alter streams, backwaters, and rivers. That assumed susceptibility for purposes of endpoint selection was subsequently analyzed and confirmed.

Ecological relevance has to do with the importance of a species or other entity to the rest of the ecosystem. Ecologically relevant endpoints may not

TABLE 14.1

Generic Ecological Assessment Endpoints (USEPA 2016c)

Entity Type	Attributes
Organisms within an assessment population	Survival, Fecundity, Growth, Kills, Gross Anomalies
Assessment population	Extirpation, Abundance, Production
Assessment community, assemblage, or ecosystem	Taxa Richness, Abundance, Production, Area, Function, Physical Structure
Special places (parks, refuges, critical habitats, etc.)	Ecological attributes that relate to the special or legally protected status

be particularly relevant to the goals of decision-makers or stakeholders, but if they are susceptible the whole ecosystem is at risk. For example, the U.S. Clean Water Act specifies "the protection and propagation of fish, shellfish, and wildlife," but not phytoplankton and other plants which are highly ecologically relevant. In this context, pollutant effects on plants are treated as an indirect cause of effects on the three legal endpoint entities.

Prior to the development of ecological risk assessment, it was not recognized that, just as toxicity tests have endpoints such as a median lethal concentration (LC_{50}), assessments must have endpoints such as the species richness of stream communities. That recognition led to the development of assessments that estimate something and not just applying expert judgment to test endpoints. In particular, extrapolation models bridge the gap between the two types of endpoints (Chapter 26), and exposure models estimate exposures of defined endpoint entities.

14.3 Human Health Assessment Endpoints

Assessment endpoints for human health are much more limited than for ecology. There is only one entity, humans, and two broad attributes: cancer and adverse non-cancer effects. Fear of cancer has been a major impetus for environmental laws, so Superfund and most other USEPA programs include the assessment endpoint of lifetime cancer risk (USEPA 2014c). However, human risks are usually expressed as frequencies of cancer or other effects in a population. Examples of non-cancer adverse effects include asthma, IQ decrement, low birth weight, and developmental defects.

Because human health risks may consider suborganismal effects that are extrapolated to health effects, adaptive effects must be distinguished from adverse effects. For example, different pathologists interpreted observed cell proliferation in thyroid glands exposed to perchlorate as adaption to the toxic effects (hypertrophy) or as a pre-cancerous state (hyperplasia). Cell proliferation was not a sufficiently defined test endpoint, because only hyperplasia was interpreted as an early stage of the assessment endpoint, thyroid cancer. However, in some cases, a precautionary approach is taken, and any cell proliferation is treated as pre-cancerous.

14.4 Organisms versus Individuals, a Source of Confusion

Traditional toxicity tests are performed on organisms, and the ecological test endpoints are attributes of organisms: survival, fecundity, and growth.

Organism-level attributes fit nicely with the assessment endpoints for human health because individual humans are valued. However, it creates confusion in ecological assessments. Critics accuse the USEPA and other agencies of treating nonhuman species like humans by protecting individuals. This attack is resolved by considering the entities and attributes of the endpoints separately. The endpoint attributes are attributes of organisms, but the entity is a population of organisms. For example, an LC_{50} from a fish test is interpreted as the concentration at which half of the fish die, not at which there is a 50% probability that an individual fish will die. Hence, the LC_{50} test endpoint may be interpreted as the concentration at which mass mortality (the attribute) would occur in an exposed population of fish (the entity), which is an ecological assessment endpoint. If the effect occurs in multiple exposed species such as fish of multiple species dying in an observed fish kill, the entity may be community level, but the attribute (death) is still organism level.

The concept of taking provides an exception to the generalization that individuals are not ecological endpoint entities. U.S. law forbids taking of threatened or endangered species, marine mammals, or bald and golden eagles. Taking means to harass, harm, pursue, hunt, shoot, wound, kill, trap, capture, or collect individuals of the protected species. Hence, organism-level attributes may be applied to individuals of these highly protected species.

14.5 Other Contexts

The assessment endpoints concept was developed for ecological risk assessments in the U.S. Other nations and other assessment contexts have applied different concepts and terminology to describe what is protected by an assessment-informed decision. For example, EFSA has guidance similar to generic assessment endpoints termed "specific protection goals options" (EFSA Scientific Committee 2016). Australian impact assessments address items from a list of "matters of national environmental significance" (Department of the Environment 2013).

14.6 Ecosystem Services

Ecosystem services are "outputs of ecological processes that contribute to human welfare or have the potential to do so in the future" (Munns et al. 2016; USEPA 2016b; Faber et al. 2019). Most environmental laws in the U.S. do not require demonstration of a benefit to humans before protecting the environment, and most ecologists and environmental advocates argue for protecting the environment for its own sake. However, the political benefits of recognizing human self-interest in environmental decisions are obvious. The influential Millennium Ecosystem Assessment defined four categories of ecosystem services—supporting, provisioning, regulating, and cultural—which have been adopted by some agencies.

Ecosystem services have three uses in environmental assessments: (1) support for ecological assessment endpoints, (2) meeting the human welfare goal, and (3) support for cost/benefit analysis.

14.6.1 Support for Ecological Assessment Endpoints

Even when ecological assessment endpoints are well supported by law, policy, and precedent, a decision-maker may be disinclined to act to protect entities that they do not understand (e.g., cryptogamic crusts) or that they perceive as useless (e.g., bugs in the mud). Similarly, stakeholders may not support protective actions if they do not see a benefit. In such cases, by citing ecosystem services, you can augment the support for a conventional ecological assessment endpoint. For aquatic insects, the obvious service is providing food for fish that humans catch and eat, but there are many more (Suter and Cormier 2015b). For example, nutrients are lost to watersheds when they leach into streams. However, aquatic insects restore nutrients to soil when they emerge and die on land. A conceptual model can show the direct benefits to people of that service (Figure 14.1).

The numerous services of nature are best conceived as providing benefits associated with human values (Table 14.2). The endpoints that provide consumptive and recreational values are the only ones that are routinely quantified. Other services are typically described and used as qualitative support. They help to answer the "so what" question.

14.6.2 Endpoints for the Human Welfare Goal

Some laws and regulations, such as the U.S. Clean Air Act, set protecting human welfare as a goal. Human welfare is a broad concept that could encompass nutrition, health, comfort, self-actualization, security, cultural identification, and other human conditions. I witnessed this breadth of welfare effects in Bristol Bay where the anxiety of the Yupik people was unmistakable. More than health or nutrition effects, they feared that the Pebble Mine would destroy their salmon-based culture. With respect to U.S. environmental laws, welfare refers to properties of the environment other than toxic effects on health that are relevant to humans, and it is assessed by ecologists. The assessment endpoints for welfare in environmental assessments may be defined as products of ecosystem services as in Figure 14.1. An example is the Integrated Science Assessment for ozone, which estimated the loss of forest production (an ecosystem service) as an endpoint for human welfare risks at alternative ozone standards (USEPA 2014e).

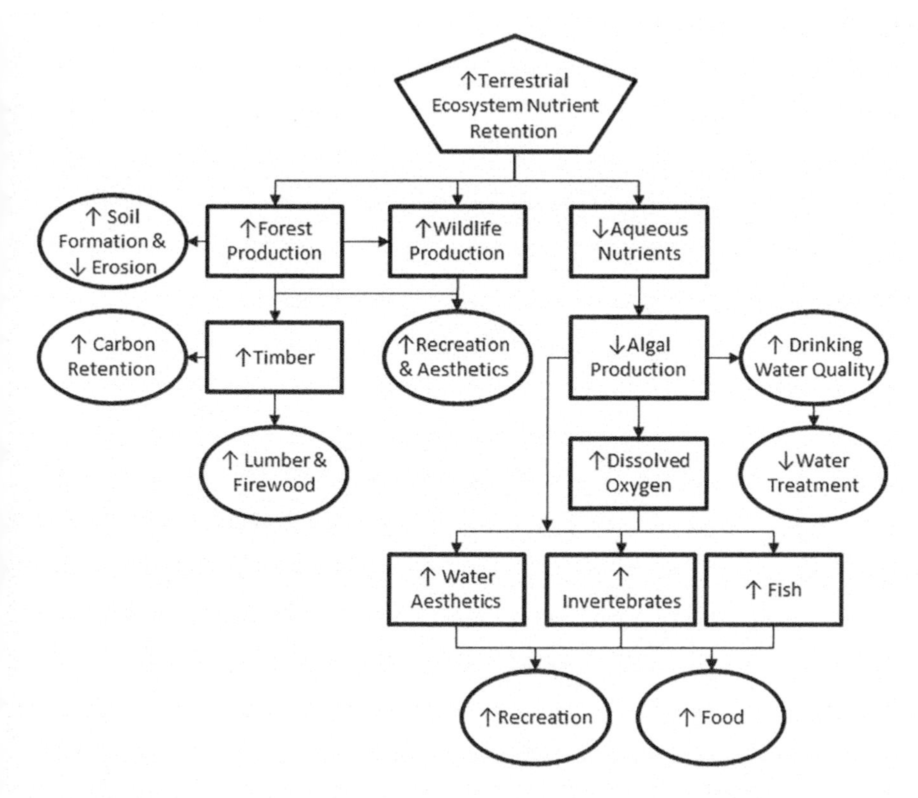

FIGURE 14.1
A conceptual model of the benefits of terrestrial nutrient retention by aquatic insects, an ecosystem service (Suter and Cormier 2015b). Ovals are benefits, rectangles are steps in the processes that generate those benefits, and arrows in boxes indicate whether the variable will increase or decrease.

14.6.3 Support for Economic Analyses

Since a 1981 executive order from Ronald Reagan, major Federal actions, including USEPA regulations or standards, must be economically justified. This book does not address economic assessment methods, but identification of ecosystem services and estimation of the potential loss of such services can facilitate the economic analysis. Assessment of ecosystem services endpoints can provide a bridge between effects on ecological endpoints and monetary benefits of protection. However, there may not be good correspondence between ecological assessment endpoints and the monetary benefits of a protective action. Ecological endpoints are focused on sensitive entities and attributes, including those that are supported by policy and precedent and those that have specific legal protections such as endangered species. They do not necessarily have the most monetary value. To assure complete consideration of the benefits of environmental protection, it seems appropriate to include all effects, or at least those that have obvious economic value, to get total benefits, not just the

TABLE 14.2

Examples of Ecosystem Services Associated with Human Values (USEPA 2016c)

Environmental Value Categories	Definition	Example Ecosystem Services Endpoints
Consumptive:	Value of commodities produced by environment	Mass of wood produced Volume of clean water provided Mass of fish produced
Informational	Value of environment as models for anthropogenic structures, chemicals, and processes	Novel chemical or physical structures potentially identified (unquantified)
Functional/structural	Value of ecological functions and structures	Quantity of carbon sequestered Extent of flooding avoided
Recreational	Value of recreational opportunities	Number of visitor day opportunities
Educational	Value of educational opportunities	Number of visitor day opportunities
Option	Value to future generations from environmental preservation	Quantity available in the future
Existence	Nonuse value of environment	Spiritual, inspirational, or aesthetic importance

most sensitive. This is reasonable for human health but is difficult for ecology because there are so many potential ecological endpoints.

A relatively simple example of the ecological economics problem is the Bristol Bay regional assessment. The primary ecological assessment endpoint was abundance and production of salmon. The services of salmon include human consumption and recreation. Both of those services were readily quantified and monetized in terms of the contribution of the salmon industry and recreational fishing to the economy. However, the cultural value of the salmon to the indigenous people of Bristol Bay watershed cannot be monetized by market economics. Calling it a cultural ecosystem service may help to raise the issue, but cultural effects could not be quantified much less monetized. Trying to include all effects on all species and products of the loss of stream and wetland habitat and of salmon production such as reduced abundance and production of salmon predators and scavengers and reduced terrestrial and aquatic primary and secondary production, was simply not feasible. Fortunately, the direct monetary benefits of salmon in the watershed were huge.

Cost/benefit analysis and other economic techniques are outside of my area of expertise, and the economic analyses that I have known were, in my opinion, highly incomplete. Decisions must be reasonable, and clearly, the costs of an action can be too large relative to the environmental benefit. However, few ecological benefits can be monetized, so economic analyses are commonly driven by loss of human lives, defined by the value of statistical life or value of mortality risk (NCEE 2014).

15

Precaution and Protection

Environmental laws require protection, safety, or the absence of effects, which requires taking precautions in the face of uncertainty. The precautionary principle has been written into some laws, but the definitions are vague and the interpretation is often difficult. Some definitions require prohibiting an action when effects are uncertain. Such definitions would stop all progress and they have in fact been used to block genetically modified organisms in many cases. More practical definitions may be generalized as follows: if an agent or action is potentially dangerous and you are uncertain about the danger, take protective measures. The precautionary principle may be related to the engineering practice of routinely designing structures and devices to be stronger than calculations require. It is also analogous to the physician's admonition, first do no harm. Its opposite is the principle that a chemical or other agent is innocent until proven guilty, which has prevailed under the U.S. Toxic Substances Control Act (TOSCA).

The precautionary principle began with Principle 15 of the Rio Declaration on Environment and Development of 1992:

> In order to protect the environment, the precautionary approach shall be widely applied by States according to their capabilities. Where there are threats of serious or irreversible damage, lack of full scientific certainty shall not be used as a reason for postponing cost-effective measures to prevent environmental degradation.

Full scientific certainty never exists, so this statement is in effect a warning against the "doubt is our product" approach to preventing protective actions (Chapter 20). However, the requirement that the threats be "of serious or irreversible damage" seems to negate the precautionary nature of the rest of the statement by requiring some evidence of substantial risks. Finally, the term "cost-effective" seems to require risk and economic assessments demonstrating acceptable costs given the environmental risks. These ambiguities are typical of other statements of the precautionary principle since the Rio Declaration.

The precautionary principle was made a fundamental part of European Union health and environmental protection by the Treaty of Maastricht in 1993. However, its application to chemical permitting, water and food quality,

DOI: 10.1201/9781003156307-17

and other issues has been too vague and various to summarize here. It is a noble sentiment, but hard to apply.

The U.S. has not officially adopted the precautionary principle. In fact, there have been efforts to prohibit the precautionary principle in the U.S. However, in practice, environmental laws are usually implemented in a precautionary manner in the sense of taking actions to reasonably ensure protection.

16

Levels of Organization

Life is nested biomes all the way down.

—*Merlin Sheldrake, Tangled Life*

We scientists, and especially ecologists, are fascinated by the hierarchical organization of the natural world. We convey this by defining issues in terms of levels of organization, which can be both revealing and misleading.

The levels of biological organization considered by ecological assessors are suborganism, organism, population, community, ecosystem, and biosphere. Human health assessments are concerned with the health of human organisms within a population (i.e., the frequency of an effect in an exposed population). Many authors have argued that the highest levels of organization are the most relevant and should be preferred as endpoint entities. In fact, relevance is a function of the case. Ecosystem endpoints such as primary production and nutrient turnover rates are seldom used because functional redundancy makes them relatively resistant to contaminants. Similarly, community attributes such as species richness tend to be more resistant than population attributes such as abundance. Further, higher level entities generally occupy larger areas than lower levels. Therefore, effects of local contamination or disturbance may be severe but are diluted out in uncontaminated areas occupied by a larger endpoint entity. The choice of level of organization depends on what is sensitive and valued (Chapter 14).

16.1 The Levels

Suborganism—Suborganismal levels include everything from molecules to organ systems. Among the USEPA's generic ecological assessment endpoints, suborganismal assessment endpoints are limited to gross anomalies (deformities, tumors, and lesions) (USEPA 2016c). Examples include the water quality criteria for tributyl tin and selenium, which are based on deformities in mollusks and fish, respectively. Human health assessments include suborganismal endpoints that have implications for health such as micromorphological changes in tissues (e.g., necrosis, metaplasia, or atrophy) that could result in severe dysfunction. Note that effects like cranial deformities of fish

DOI: 10.1201/9781003156307-18

or liver dysfunction in humans are endpoints, because of their implications for a higher level, the organism as a whole. It is best in those cases to define the endpoint entity as the organisms of concern and the endpoint attribute as the suborganismal structure or process that is potentially affected. Suborganismal effects have uses beyond assessment endpoints. They are used primarily to characterize modes of action and as biomarkers of specific causes in field studies. Similarly, molecular and cellular test data are increasingly common and important in hazard identification for toxicological assessments (Chapter 12).

Organisms—Organisms are, with the exception of some colonial species like corals, readily identified. They are the most common and seemingly the most natural level of organization for environmental toxicology and assessment. In general, ecological assessors, environmental advocates, and decision-makers are careful to state that they are not protecting individual fish, trees, or other nonhuman organisms. However, there are exceptions (Chapter 14). For example, in the U.S., to protect the national emblem, the Bald and Golden Eagle Protection Act prohibits "taking" bald or golden eagles, including their parts, nests, or eggs. Most other organisms are not so protected, but that may change. Some animal rights advocates have discovered the environment and advocate protection of all vertebrate animals as individuals.

Protection of individual animals can be problematical in practice. During the Superfund cleanup of the Oak Ridge Reservation, a ground hog (a species of marmot) was found to have burrowed into a low-level radioactive waste trench. A representative of the State of Tennessee proposed that we assess the radiation risk to that individual even though it was not representative of any population and had no protected status. While we worked to resolve that point, the site manager, who was a practical engineer and unaware of the State's issue, had the animal shot and the burrow filled to restore isolation of the waste.

Populations—In theory, a population can be defined as an interbreeding group of organisms. However, you can seldom delimit such a group in the field. In practice, we must define an assessment population that is relevant to our assessment. Populations are common endpoint entities because they are frequently of interest to decision-makers and the public. The valued populations include fisheries, wildlife harvested by hunters, wild food species such as mushrooms, and remnant populations of endangered species. Populations are also the presumed endpoint entities for most ecotoxicological assessments even though the endpoint attribute is, as explained above, usually organism level.

We assess populations in three distinct ways. First, and most commonly, studies of effects on organism attributes, particularly survival, growth, and reproduction, are interpreted as effects on the members of a population (Chapter 14). Those attributes are included in ecotoxicity tests because they have implications for population survival and growth. However, they are treated separately, and the test endpoint is the most sensitive response. For example, a 10% effective concentration (EC_{10}) for fecundity may be the test endpoint because it is lower than the EC_{10}s for mortality or growth. Second, the test attributes can be used to parameterize a population model and estimate a population variable such as the intrinsic rate of increase. This makes sense to me but is rare. Third, effects on actual populations may be observed in the field or tested, for small populations, in microcosms and mesocosms (Chapter 21).

Human health endpoints are usually expressed as the frequency of some organism-level attribute (e.g., cancer) in a population (e.g., farm workers). Therefore, although the intent is to protect human individuals, human health assessments analyze effects from exposure of a population. However, it may be an exposed population such as the farm workers or a susceptible population such as pregnant women rather than a biological population.

Communities—In theory, a community can be defined as an interacting group of populations. As with populations, communities are seldom naturally delimited in the real world. In addition, a delimited community for some types of organisms will not encompass others. Therefore, we define endpoint communities as the set of populations occupying an area relevant to an assessment.

Ecosystems— Communities plus their physical environment constitute ecosystems. Hence, ecosystem endpoints focus on physical structure and function. Common examples include the area of wetland lost to drainage or filling, the acidification of a stream by mine drainage, and the eutrophication of a lake receiving fertilizer runoff. Some ecosystem processes such as organic matter decomposition rates in soils have been advocated and tests have been developed, but they are seldom used. This is in part because functional redundancy makes them insensitive. If one cellulolytic microbial taxon is lost, there are plenty of other microbes that can perform that function. An endpoint ecosystem is defined by the area within which the ecosystem endpoint can be characterized and which is subject to physical or chemical alteration.

Biosphere—The biosphere is the portion of the planet that supports life. It is different from even a very large ecosystem in that it is all-inclusive and is effectively closed with respect to organisms and other matter. Potential endpoint attributes include planetary biodiversity, biogeochemical cycling processes, and of course, climatic attributes. The few assessments at this scale include the IPCC assessments of climate change and its implications for the habitability of the planet. Another example is the International Union for the Conservation of Nature's Red List of Threatened Species.

16.2 Levels of Organization of Assessment Endpoints

Assessment endpoints are defined by an entity and an attribute of that entity that should be protected (Chapter 14). Much confusion has occurred due to failure to recognize that the entity and attribute may not be defined at the same

TABLE 16.1

Examples of Assessment Endpoints for Human and Ecological Risk Assessments

Entities	Human Health Assessment	Ecological Assessment
Organism-level attributes		
An individual organism	Probability of death or injury (e.g., risk to the maximally exposed individual)	Probability of death or injury (e.g., risk to an individual of an endangered species); Seldom used
A population of organisms	Frequency of death or injury, numbers dying or injured	Frequency of mortality or gross anomalies, average reduction in growth or fecundity
Population-level attributes		
An individual population	Not used	Extirpation, production, or abundance
A set of populations	Not used	Extinction rate or regional loss of production; Seldom used

Source: Adapted from Suter, Norton, and Fairbrother (2005).

level of biological organization (Suter, Norton, and Fairbrother 2005). In particular, the most common levels of assessment endpoints are organism and population, but these terms may refer to different combinations of entity and attribute (Table 16.1). In human health assessments, the goal is to protect humans, including sensitive individuals. However, the conventional endpoint attributes are frequencies of dichotomous organism-level states: alive or dead, cancerous or not, deformed or not, etc. Because frequencies imply a population, human health assessors refer to these as population endpoints. So, the assessment endpoint entity is a human population, and the attribute is frequency of a condition. Continuous response variables are also used: birth weight, growth rate, memory test scores, etc., but they are often converted to frequencies within a population of dichotomous endpoints such as low birth weight versus normal birth weight.

It is important to be aware when communicating results of the implications of effects at one level for other levels that may be better understood by your audience. For example, if you conclude that species richness was reduced at a contaminated site, also state which species or higher taxa were lost. For human health, explain the implications of suborganismal effects such as liver pathology on a person's health and longevity.

16.3 Delimiting Endpoint Entities

Your assessment may estimate an absolute loss (e.g., a spill killed 112 fish or a highway will destroy a hectare of oak-hickory forest), but you may also provide perspective. You may determine the magnitude of effects relative to the size of the affected population, community, or ecosystem. For that, you must delimit the system. Otherwise, you cannot say whether a spill that killed 112 fish extirpated the population or killed only a few percent of the population. Although organisms are, with some exceptions like corals, easily delimited, defining the limits of a population or community is generally difficult. Animals move, plants disperse seeds, and habitats may extend for thousands of square kilometers. Populations may form intermittently and use different habitats in different seasons. An example is the flock of migrating snow geese that landed in the flooded Berkeley pit in Montana and thousands died. Populations are defined as interbreeding, but that unfortunate flock could be considered a migratory population.

The solution to the delimitation problem is to be pragmatic and define populations and communities to suite the assessment. An assessment population, community, or ecosystem is a system occupying an area that has been defined as relevant to an assessment (USEPA 2016c). It may be a stream, watershed, contaminated site, an area treated with a pesticide, or something else. It may be actual or hypothetical, but it should be reasonable. It is unreasonable, for example, to define a one-hectare brown bear population. At the other end of the size spectrum, concerns about extremely large-scale endpoints such as predicted continental or global loss of biodiversity may be responsive to the assessment's goals but with questionable reliability.

17

Confidence and Uncertainty in Risk Characterization

The question is not "do we know everything?"
It is "do we know enough?"
or "how can we best make a decision using what we know?"

—*Sense About Science (2013)*

You may think of confidence and uncertainty as inverse concepts, but it is not that simple. We can distinguish these qualities from their uses in the USEPA's risk characterization handbook (Science Policy Council 2000). The IPCC's guidance on treatment of uncertainty also requires reporting confidence and uncertainty but with somewhat different definitions (Mastrandrea et al. 2010). The National Acid Precipitation (NAPAP) assessment plan called for defining confidence levels and unavoidable scientific uncertainties. NAPAP management recognized that, from a policy and action perspective, an emphasis on uncertainty was a recipe for paralysis (Oppenheimer et al. 2019). The distinction between confidence and uncertainty is applicable to all types of environmental assessment. The reader should be aware that the distinction is a best practice, but it is not universal. Many assessments report only uncertainties with their results.

17.1 Confidence

"Your major responsibility as a risk assessor is to communicate your key risk findings and conclusions and your confidence in them in the risk characterization section of your assessment" (Science Policy Council 2000). As that quote suggests, you should express your confidence in your assessment results, so that managers know how much confidence they can have in a decision based on your results. Expressions of confidence in risk assessments may be quantitative, based on the relative magnitude of the exposure level and effects level, the margin of exposure (MOE=benchmark value/exposure level). If those values are not close (MOE is much larger or smaller than 1), you can be confident in the effect or absence of effect. Note that the MOE is

DOI: 10.1201/9781003156307-19

the inverse of the risk quotient (RQ). However, they are used differently. RQs are decision criteria. If RQ >1 or any other preset criterion, the risk is considered at least potentially significant. MOEs are used to express confidence that the exposure will or will not exceed the benchmark.

A judgment of confidence also requires confidence in the evidence behind the numbers. Ideally, confidence in the evidence is determined by weighing the evidence (Chapters 24, 25). The IPCC defines confidence as "confidence in the validity of a finding, based on the type, amount, quality and consistency of evidence (e.g., mechanistic understanding, theory, data, models, expert judgment) and the agreement" (Mastrandrea et al. 2010). That is, a weighty body of evidence should impart high confidence in the evidence and hence confidence in the result.

17.2 Uncertainty

"Uncertainty represents lack of knowledge about factors such as adverse effects or contaminant levels which may be reduced with additional study" (Science Policy Council 2000). Often, these known unknowns are expressed as lists of missing information such as fish chronic toxicity, biodegradation rate, or concentration in food items. The uncertainties may also concern parameters or forms of models. Inclusion in uncertainty lists is usually a matter of judgment guided by knowledge of the information that would usually be included in an assessment. If uncertainties are quantified, the estimates are products of subjective judgments and therefore are considered to be degrees of belief (USEPA 2014d).

Uncertainty has an insidious aspect. You can always think of things that are unknown and might influence the results of an assessment. This aspect of uncertainty keeps scientists in business. There is always something more to be studied which can be the topic of a grant or contract proposal. It is also the basis for the "doubt is our product" strategy for defending products from regulation. Beginning with tobacco companies, industries have argued that uncertainties prohibit drawing any conclusions concerning risks, which in turn prevents any regulatory actions (Michaels 2008; Oreskes and Conway 2010). Similarly, conservative governments use uncertainty as an excuse for inaction. The Reagan administration emphasized the uncertainty concerning acid deposition, the Thatcher administration emphasized uncertainty concerning stratospheric ozone depletion, and the G.W. Bush administration emphasized uncertainty concerning climate change. On the other side, chemophobes argue that there is too much uncertainty to allow the use of synthetic chemicals. The emphasis on confidence in risk characterization can mitigate an insidious emphasis on uncertainty. When assessing uncertainty, keep in mind that you can never know everything concerning a hazard, but you can assess what is important to know and what is reasonably knowable.

17.3 Unifying Confidence and Uncertainty

"To ensure transparency, risk characterizations should include a statement of confidence in the assessment that identifies all major uncertainties along with comment on their influence on the assessment" (Science Policy Council 2000). In other words, reveal how sure you are of what you know (confidence) and reveal what you do not know (uncertainties). A quantitative example from a health assessment case is:

> The MOE estimates for drinking water exposure ranged from 106 to 107. Because these MOEs are so large, there is a high level of confidence that no appreciable concern exists due to releases to surface water. Although uncertainties associated with the duration of release of generic ketone to water could result in an increase in the surface water concentration by a factor of 37, the estimate would have to increase by a factor of 1,000 to change the MOE enough to influence the level of concern.
>
> *(Science Policy Council 2000)*

17.4 Analysis of Confidence

In risk characterizations, you may treat both confidence and uncertainty as either qualities (e.g., low, medium, high) or quantities. The examples of quantification of confidence above are MOEs. In my experience, MOEs of more than two or three orders of magnitude (100–1,000×) instill reasonable confidence in a conclusion of no effect. The interpretation of an MOE depends on your qualitative confidence in your methods and data. The MOE approach is consistent with laws such as the U.S. Clean Air Act's requirement to "protect public health with an adequate margin of safety."

The best-known quantification of confidence is confidence intervals. Assessment results may be presented as a best estimate and the confidence interval within which the result occurs with x% confidence. They need not be the conventional 95% confidence interval, and it may be advantageous to present multiple intervals such as 50%, 75%, and 95% to give the decision-makers a better idea of confidence. For example, values inside a 50% confidence interval are more likely than not, which is good to know in a decision context. It is also important to remember that confidence intervals refer to confidence in inclusion of a sample statistic, usually the mean. If you want to know confidence in the inclusion of individual observations, use prediction intervals. Confidence intervals are conventionally derived using standard parametric statistics. However, there are alternatives. Resampling techniques such as bootstrapping provide confidence intervals based on the distributions of samples from the data. Bayesian confidence intervals are called credible intervals.

Another form of confidence bounds is provided by maximum and minimum estimates. These are created by estimating effects for reasonable worst and best cases. Such cases are usually derived from exposure scenarios, but they can also be based on a range of assumptions concerning effects.

You can avoid parametric distribution assumptions, Bayesian priors, and 95% bright lines by using box and whisker plots. They summarize distributions of the data and allow you or the decision-maker to judge the confidence in effects. Conventionally, they are graphic presentations of the mean, median, interquartile range and data range. An advantage relative to parametric confidence intervals is that they show skewness in the data.

Note that what is known as uncertainty analysis is actually a tool to address confidence, not uncertainty as used in the USEPA characterization guidance. For example, a Monte Carlo simulation of a mathematical model is often described as uncertainty analysis, but it is typically used to generate confidence intervals on mean or median model results.

Confidence also includes the qualitative weight of the evidence for your results. A confidence interval or MOE gives you high quantitative confidence,

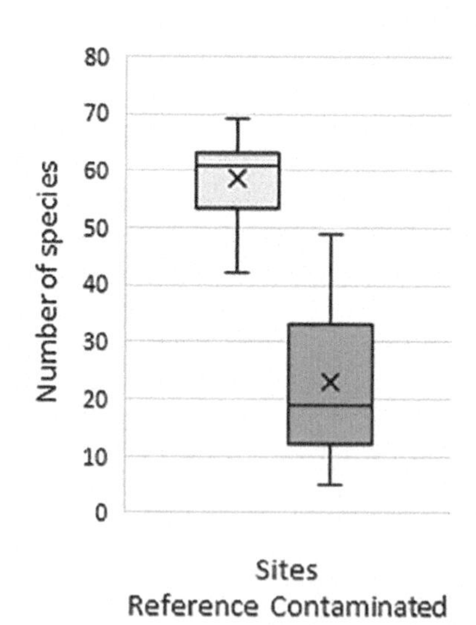

FIGURE 17.1
Box and whisker plots for the number of invertebrate species found in reference and contaminated streams. Boxes include 50% of the sites and whiskers contain the range. The line in the box is the median and the *X* is the mean. The numbers are invented for illustration.

but the evidence behind it, including qualities such as the reliability of the study design used to generate the data, also contributes to confidence. The weight of evidence is your confidence in the evidence that went into your results. A formal weight-of-evidence analysis is desirable (Chapters 24 and 25), but even a subjective judgment is useful. For example, "the evidence is directly relevant and highly reliable."

17.5 Analysis of Uncertainty

In most assessments, you disclose your uncertainties in a list containing the things that we do not know that could be important. Your list should distinguish uncertainties that are likely to influence the results or at least uncertainties that are due to the lack of information that is usually available in assessments of the type. These judgments are based on your experience and on simply adding to an uncertainties list as you perform an assessment and find that you lack certain information.

For uncertainties that could change the results, quantification of uncertainty can be useful. Analysis of the potential influence of a parameter on results is termed sensitivity analysis. The need for an uncertain parameter can be determined by performing a sensitivity analysis of changes in results from credible values of the parameter. That sensitivity can be compared to quantified confidence in the results. In the ketone example in the preceding subsection, the factor of 37 is the sensitivity to the uncertain parameter and the MOE of 1,000 is the confidence.

Often, confidence is neglected and only uncertainly is addressed. Uncertainty, in those cases, may encompass what is imperfectly known, what is unknown, and what is known but variable. Tiers of uncertainty analysis have been proposed by the WHO for human exposure (WHO 2008), by the USEPA for Superfund and National Ambient Air Quality Standards (NAAQS) (USEPA 2020b, 2001), and by the EU for pesticides and other chemicals (EU 2005; EFSA 2006). The following generalization of those documents summarizes the hierarchy of options:

Tier 0—used for routine and screening assessments, lists the uncertainties, and may include applying default uncertainty factors or conservative assumptions.

Tier 1—the lowest level of chemical- or site-specific uncertainty characterization, involves qualitative characterization of sources of uncertainty (e.g., a qualitative assessment of the general magnitude and direction of the effect of an uncertainty on risk results).

Tier 2—chemical- or site-specific deterministic quantitative analysis involving sensitivity analysis, interval-based assessment, and possibly probability bound (high- and low-end) analysis.

Tier 3—uses Monte Carlo or Bayesian probabilistic methods to characterize the influence on results of individual or combined sources of uncertainty.

In many regulatory contexts, you address uncertainties during the estimation of effects levels by applying factors (Tier 0) (Chapter 26). USEPA guidance deals with the absence of data for relevant taxa, life stages, or effects by dividing by factors of 10. The sources of uncertainty need not be reanalyzed for the risk characterization because these standard factors are assumed to be adequate and applied in the analysis phase. However, those who treat uncertainty by applying factors may still analyze and report confidence (EC 2011).

Results of statistical analyses of variables may be considered expressions of confidence or uncertainty, depending on your interpretation. For example, conventional uncertainties in species sensitivity distributions (SSDs—Chapter 27) are deviations of points from the distribution function that is fit to the species data, and I would treat that as confidence in the model rather than as uncertainty. It does not include the absence from the distribution of important taxa, the fact that species sensitivities are often not symmetrically distributed, or other fundamental unknowns that are properly termed as uncertainties in SSDs (Posthuma, Suter, and Traas 2002). Bootstrap confidence intervals on SSDs are based on resampling from the set of species, so they can be interpreted as confidence given an incomplete set of species. For an example, see Figure 11 in USEPA (2011b).

The ultimate uncertainty is unknown unknowns. For example, ozone depletion was assessed as a slow global gas-phase reaction, and the concentrations of chlorofluorocarbons and reaction rates were unknown. Then, an unknown unknown was discovered, the Antarctic ozone hole and catalysis by polar stratospheric cloud particles, which rendered prior assessments irrelevant. Similarly, early risk assessments of DDT were rendered irrelevant by the discovery of an unknown phenomenon, trophic biomagnification.

17.6 Uncertainty versus Variability

Although they are often lumped, it is generally useful to distinguish lack of knowledge from the inherent variability in the system being assessed. Lack of knowledge might include the toxicity of a chemical to birds, the soil biodegradation rate of a chemical, or the failure rate of a new treatment technology. Variability includes stream flows, wind directions, composition of municipal wastewater, and size of individuals in a species. The distinction has implications for assessments. Lack of knowledge can potentially be corrected by testing, measurement, or even modeling. Variability can be observed and characterized as a distribution, but it cannot be reduced. Variability is a property of the system, but uncertainty is a property of the observer. Results of risk assessments are often expressed as frequencies of something bad, which ideally is based on knowledge of variability but may also be based on subjective probabilities (Chapter 18). For example, variability may be treated as determining a protective value and confidence in that protection. You might conservatively estimate exposure in a receiving stream based on the tenth centile flow and derive a confidence interval on that value, all based on variability of flow.

You can go further and integrate multiple variable parameters. Exposure to the chemical in the stream is variable due to the variance in the effluent flow, the receiving stream flow, and the concentration of the chemical in the effluent. The distribution of exposure concentrations with respect to time can be calculated as the joint distribution of those parameter distributions. That calculation is appropriate because all variability is with respect to a common unit of time. The joint distribution might be used to estimate the frequency of exceedances of a water quality standard. Other uncertain variables in your exposure model such as the unmeasured partitioning coefficient of the chemical between the particulate and dissolved phases, are not included in the joint distribution with respect to temporal variance.

17.7 Making Sense

Confidence, uncertainty, and variability are often lumped together as uncertainty. Uncertainty may be expressed as qualitative scores (High, Moderate, Low) and, if quantitative, may be expressed as probabilities. However, qualitative uncertainty may be vague, and probability in itself means nothing in particular (Chapter 18). To make sense to decision-makers and other readers, you must explain what you are reporting. To do that, you need to understand the meaning of the analyses that you or your statistician are doing.

For qualitative uncertainty expressions, you should provide an explanation of each term. For example, low uncertainty may be defined as "the lack of information concerning a variable is unlikely to influence the results." For quantitative expressions, you should start with individual computations and think about what they represent. For example, as discussed above for SSDs (Section 17.5), lack of fit of a parametric model tells you about confidence in the model given the data. If you want uncertainty concerning incomplete representation of species, you should use bootstrapping to show the influence of differences in species data sets. For me, the resampling technique is closer to what I want to know than is the fit of data to a presumed true function.

17.8 Reporting Confidence and Uncertainties

The USEPA Science Policy Council (2000) chose the Office of Pesticide Program's risk characterization of the acaricide Mitec as exemplary. The quotations below represent summary statements concerning two exposure scenarios and the associated risk. Note that the critical information is the confidence in the results. Also note that confidence comes primarily from weighing the reliability of the evidence.

Dietary exposure:

> OPP has high confidence in the dietary risk estimates. Two reliable sources of residue data (the market basket survey and USDA's PDP) are in close agreement, and OPP used the food consumption data from the 1989–1992 USDA CSFII survey, which is the latest survey available.

Occupational exposure:

> OPP has low overall confidence in many of the occupational exposure scenarios due to a low number of replicates or poor data quality. Therefore, the occupational exposure estimates should be considered preliminary.

Dietary risk:

> The refined total dietary cancer risk for Mitec in the U.S. population was 1.6×10^{-5}. OPP has very high confidence in this estimate because it is based on actual residue data, the latest USDA food consumption data, a Q1* based on a rare tumor type, and a scientifically defensible dose-response extrapolation.

Uncertainty was discussed separately largely in terms of transparently admitting lack of knowledge. "Identify those scientific uncertainties that if reduced ... would make a real impact on the risk assessment" (Science Policy Council 2000). That is, list information that you judge, based on experience, could be potentially influential, and if feasible, estimate the magnitude of uncertainty.

Although the quoted examples are old, they are still USEPA policy and in fact are still state-of-practice risk characterizations. However, you would do well to apply a little more consistency and formality. The weight-of-evidence guidelines recommend that characterization of confidence for a quantitative assessment result should include both a statistical characterization such as confidence limits and a qualitative characterization of the weight of evidence (USEPA 2016e; Suter, Cormier, and Barron 2017). The statement above concerning confidence in the dietary data is an unstructured narrative weight-of-evidence conclusion.

17.9 Summary

In risk characterization, you provide the results, given the data and assumptions. The users of your assessment also need to know how confident you are and how confident they can be if they use your results. The unknowns are the things that are left out. You should be transparent about the uncertainties that come from not knowing those things. With greater confidence, the concern about uncertainties is diminished. The call to characterize both confidence and uncertainty arose in risk assessment, but is also applicable to other types of assessments.

When analyzing confidence and uncertainty, think carefully about what you are doing and how it should be interpreted. Do not just apply a handy technique and report a result.

18

Probabilistic Assessment

People are very bad when reasoning about risks. We make a lot of mistakes when we use probabilities.

— *Ian Hacking (2001)*

When an ORNL team started to develop ecological risk assessment for the USEPA, we thought it should be probabilistic. We developed methods for estimating organism, population, and ecosystem effects as probabilities and proudly presented them to the Agency (Barnthouse and Suter 1986; Barnthouse et al. 1982). They bombed. They were accepted by peer-reviewed journals (Barnthouse et al. 1987; O'Neill et al. 1982; Suter, Vaughan, and Gardner 1983) and by the Agency's external review panel, but they were not used by the Agency. The probabilistic risk assessment methods developed by the Ecological Committee on FIFRA Risk Assessment Methods (ECOFRAM) were more advanced, but they met a similar fate in 1999 (USEPA 2020).

To this day, some environmental risk assessors contend that risk assessment results must be probabilities. However, experience has clearly shown that risk assessments not only need not be probabilistic but also that probabilistic results are often unclear. Decision-makers do not want to make a decision based on a finding that the probability of impairment is 0.43, and in many regulatory contexts, they cannot. For example, fining an emitter for exceeding a criterion cannot be based on a probability that the criterion was exceeded.

Our inspiration for probabilistic assessment was the WASH-1400, reactor safety study. It used fault tree/event tree analysis of component failures to estimate the annual probability of a reactor core meltdown and subsequent deaths. The bottom line was an annual probability of death from a 100-reactor industry of 2×10^{-10}. Such engineering analyses defined probabilistic risk assessment at the time. However, it has proven to be irrelevant to actual reactor accidents since then, which resulted from partial failure of a component (Three Mile Island), human error (Three Mile Island and Chernobyl), and designing for natural events less extreme than earthquakes and tsunamis in the geological record (Fukushima Daiichi). The probabilistic analyses in WASH-1400 seemed impressive, but because they were limited to probabilities that could be calculated from rates of complete component failures, they proved to be worthless as predictors.

DOI: 10.1201/9781003156307-20

18.1 Defining Terms

What are probabilities? You might think that there is an agreed-upon statistical answer. Rather, defining probability is at the core of the ongoing "statistics wars" (Clayton 2021). A minimal definition is, probabilities are numbers on a 0–1 scale, where 0 represents impossibility and 1 represents inevitability. Probability may be mathematically interpreted as the common scale that provides the basis for axiomatic probability theory. I believe that you will agree that definition may be true, but it is not of much use. The historically foundational definition is that probabilities express the chance of an outcome from a randomness generating device for gambling such as dice, cards, or coins. That probability, defined as chance or propensity, is also of not much use to us. In scientific practice, the probability scale is a mathematical abstraction that is used to represent either frequencies or degrees of belief.

Frequencies may be natural frequencies which are the number of occurrences within a population or a unit of time (12 out of 36 mice died). They may be converted to normalized frequency (333 out of 1,000) for ease of comparison to other normalized frequencies. Either type of frequency may be expressed as a probability. The probability that a member of the mouse population is dead equals the proportion dead, 0.33. They appear in environmental assessments primarily as frequencies of undesired events such as oil tanker spills and water treatment failures or as frequencies of effects such as cancers in exposed human populations. They may be estimated by observed frequencies such as the frequency of leaks per oil pipeline-mile compiled by the U.S. Department of Transportation. Alternatively, they may be modeled as the proportion affected in a dose–response model of a rat cancer test. Proportions, as in the proportion of a population with cancer, are 0–1 normalized frequencies within members of a group or set.

Degrees of belief are much less common than frequencies in probabilistic expressions of environmental risk. Or at least, the subjectivism is unacknowledged. Subjective probabilities are described as equivalent to the odds that you would require to take a bet that, for example, a tanker will wreck and spill oil. An expert on tanker safety might consider the record of the tanker fleet, advances in tanker design, changes in operation practices, and the tankers usual route to come up with a subjective probability that the tanker will wreck in the next year. Most subjectivist probabilities are expert judgments of that sort, not Bayesian conditional probabilities. Most degrees of belief in probabilistic assessments are beliefs in some sort of frequency. However, some are just beliefs. For example, if I had been asked to express the conclusion of the causal assessment of the decline in kit foxes on Elk Hills as a probability, I might have reluctantly said 0.8. It was a one-time event, with no relevant historical frequencies so that is a subjective probability.

Frequencies and degrees of belief are associated with frequentist statistics and Bayesian statistics, respectively, but they are not necessarily tied to them. The discussions above of probabilities as frequencies and as degrees of belief make no reference to those statistical schools which have waged the statistics wars. Both statistical schools have major limitations that have led some statisticians to call for a new conceptual basis for statistics. Frequentist analyses rely on a hypothetical infinite replication of a random process to define a distribution. They can calculate the probability of data given a hypothesis, but not the probability of a hypothesis given data. On the other hand, Bayesian statistics rely on a subjective prior probability and, because they deal in subjective degrees of belief, they are immune to validation or refutation (Box 18.1).

BOX 18.1 WHAT DOES BAYESIAN MEAN?

Bayesian methods are popular now and are mentioned in Chapters 17, 18, 23, and 24. However, many of us were not trained as Bayesians so you may need some introduction.

Bayesian refers to Reverend Bayes, a 18th century English cleric, who was also a statistician and philosopher.

Bayes theory is a formula to estimate the probability of a hypothesis given data. The formula multiplies the probability of your hypothesis prior to obtaining the data times the probability of the data given the hypothesis, divided by the probability of the data. If you use this formula, you are a Bayesian.

The prior probability is your degree of belief which could be based on some information that was available to you before you gathered data. For example, the probability of two in a throw of a single die is 1/6=0.167. You have not thrown that die thousands of times to check for a bias; you just believe it is fair. In environmental assessment, you often will not have such a good basis for your prior belief. Defining the prior is the primary weakness in Bayesian practices. However, any Bayesian statistics text will present cases in which conditioning on a prior improves the analysis.

If you have no prior belief, the common practice is to assign an uninformative prior. That is, all possible hypotheses are equally probable. Uniform prior probabilities give the wrong answer unless you really have reason to believe that an effect of your treatment is no more probable than no effect (Efron 2013). If you really believed that, you would probably not conduct the experiment.

The process of applying the likelihood of your data to your prior belief is termed updating. If you have a time series such as annual surveys of relative abundances of fish species, after the first year, you have prior estimates to update.

The subjective priors are not a problem for Bayesians because they believe that probabilities are inherently subjective. Some practitioners believe that subjectivism, not Bayes' formula, is the essence of Bayesian statistics. They contend that probabilities cannot be verified because they are beliefs (Clayton 2021). I have seen cases in which some or all of the probabilities in Bayesian networks (Chapter 23) are just expert judgments and are not derived from data.

The primary advantage of Bayesian statistics, according to Bayesians, is that they derive the probability of the hypothesis given the data rather than the probability of the data given a hypothesis. The latter is referred to as the likelihood of the data. Somewhat ironically, likelihoods are what Bayesians derive from their data. It is the prior belief that turns a likelihood into the probability of a hypothesis. Distinguishing likelihoods and probabilities can be important in interpreting diagnostic tests, forensic evidence, or other such uses in which base rates are used as priors.

Some Bayesians are as reliant on data as frequentists. For example, Nate Silver uses Bayesian analysis for his 538 website. His priors for political data analyses are prior polling results and he updates them with new polling results. In 2016, he predicted that Donald Trump had only a 0.286 probability of defeating Hilary Clinton. That would seem to have been a bad prediction, but Silver pointed out that 0.286 is not zero so he was not wrong. Note that the polling organizations, from which Silver got his data, reported their results (frequencies of responses with sampling margin of error) and did not claim to calculate a probability of the electoral outcome. The difference is that Silver used Bayes' theorem to calculate the probability of Trump losing given the data, which constitutes a degree of belief in a prediction.

Probabilistic assessments are those that express results as probabilities. They answer the question, what is the probability of a bad effect? It may be an endpoint toxic effect, but more often it is the probability of a precursor event such as a pipeline failure or a permanent stream changing to intermittent following a land-use change. Estimation and reporting of confidence and uncertainties concerning a result also may employ probabilities (Chapter 17), but that is a different issue from estimating a probabilistic risk.

18.2 Uses and Methods

Most probabilistic analyses in the USEPA's assessments are Monte Carlo analyses of exposure models. An example is the assessment of human acute dietary risk from pesticides (USEPA 2000a). The Dietary Exposure Evaluation Model (DEEM) calculates dietary doses by combining the frequency distributions of consumption rates of foods with distributions of pesticide residue values in those foods. The Monte Carlo procedure samples from those distributions, calculates and saves a dose value, and repeats that process many times to generate a frequency distribution of dose levels. The 99.9th centile of that distribution is derived as an upper bound exposure which would be exceeded with a 10^{-3} probability. If that value is equal to or less than the population adjusted dose, a safe benchmark value, it is considered to be below the threshold of concern. Hence, although the method uses probabilities, the results of the assessment are conventionally deterministic and dichotomous. Had the USEPA wanted a probabilistic result, they could have derived the frequency of unsafe dietary exposures to the pesticide among members of the U.S. population and expressed it as a probability. Instead, USEPA's analysis derives a standard upper bound dietary dose to compare to the bright-line safe dose. It is a quotient method (Chapter 8) with the numerator exposure value derived using a probabilistic analysis.

Probabilistic methods like Monte Carlo analysis are often promoted as a cure for excessive conservatism. If, for example, you are assessing the human health risk from use of a pesticide on strawberries, you might use a conservative, deterministic, worst-case approach to model the exposure of a maximally exposed individual. An extreme upper end strawberry consumer may eat strawberries every day, those berries have been recently treated with the pesticide, and he never washes them. The probabilistic alternative, as described above, is a Monte Carlo analysis that repeatedly samples from the frequency distributions (scaled as probabilities) of each variable model parameter rather than assuming an upper bound value. Even the 99.9th centile of the resulting distribution of doses would still be lower than the conservative deterministic result. For that reason, industries have long advocated Monte Carlo analysis in place of conservative assumptions.

Another common use of probabilistic analysis is assessments of potential system failures. Engineers have developed fault tree, event tree, and other analyses for assessing failures of nuclear power plants, aircraft, etc., as in the WASH 1400 assessment described above. They rely on knowing the physical structure of the system and the probabilities of failure of components. Assessing failures in environmental assessments is seldom so straight-forward. When engineering analyses are not possible, the frequencies of past failures of equivalent systems can be determined. When assessing the risk of failure of earthen tailings dams for the Bristol Bay assessment, we found that three reviews of such failures estimated failure rates of roughly 1 per 2,000 mine-years. However, none of these failed dams were as large as those for the proposed Pebble mine, and the mining company insisted that large modern dams never fail. Therefore, we used Alaska's range of engineering design targets of 1 failure in 10,000 to 1,000,000 mine-years, in place of the history of failures. Not long after we issued our assessment, the large modern Mount Polley mine tailings dam in British Columbia failed catastrophically. That suggested that we should have stuck with the empirical frequencies. The same lesson could be drawn from the space shuttle failures. In the case of the Challenger space shuttle disaster, the engineering estimate of risk of catastrophic failure was 10^{-2} per launch, but NASA management insisted that it could not be higher than 10^{-5} (Richard Feynman's appendix to the Rogers Commission report). When the space shuttle program ended with two disasters, the actual frequency converted to a probability was 1.48×10^{-2} per launch. That was approximately the same as the failure rate in prior NASA programs. Never dismiss empiricism.

The essential and sometimes overlooked step in probabilistic assessment is defining what the probabilities represent. It is nearly always some sort of frequency, either empirical frequencies or degrees of belief about frequencies. In any case, it is essential to clearly identify the frequency of what with respect to what. For example, the "what" for earthen dam failures might include cracks that require repair, partial failures resulting in spillage, or catastrophic failures. The "with respect to what" may be per facility year, per dam year, per dam design lifetime, or the time until the next continental glaciation. They can be conditional, such as frequency of dam failures per year given age greater than 100 years. An example from an epidemiological study is the instantaneous (at the time of the survey) frequency of cancer in a population given agricultural work for at least the previous 10 years. That frequency of cancer may be translated to a probability of cancer in an agricultural worker if there is some reason to do so. Having defined the frequency, ask what variable parameters contribute to it. The answer might be a simple conditional probability or can become complex with Monte Carlo simulations of mathematical models or Bayesian networks. In the human daily dietary dose estimation discussed above, the probability derived by the Monte Carlo analysis represents the frequency of pesticide doses (mg/kg/day) with respect to humans in the U.S. population. The analysis is concordant because all distributions are defined with respect to variability in the parameters of individual daily dose, not degrees of belief, temporal variance, or other possible probabilities. If you are not careful in defining probabilities, you might calculate 1% of a lethal dose each day and think it is a lethal dose on 1% of days (example suggested by Scott Ferson, U. Liverpool).

It is also necessary to decide how results should be expressed as frequencies or probabilities. Frequencies are the real-world phenomenon and are often clearer than probabilities. For example, species sensitivity distributions (SSDs, Chapter 27) are understood to be models of the proportion (i.e., normalized frequency) of a biotic community that is affected. However, some assessors express the results of SSDs as probabilities but do not state probabilities of what (Suter 1998; Posthuma, Suter, and Traas 2002). They could be probabilities that a particular species or a random species is affected, but that is not how SSDs are used. Results from SSDs are naturally and clearly expressed as proportions. Even undesired events which seem naturally to be expressed as probabilities may be more clearly expressed as frequencies. It seems natural to say that the probability that a dam will fail is 10^{-5}, but because frequencies are real, they lead to asking clarifying questions as discussed above. Finally, frequencies are often more useful than the equivalent probabilities. For example, one might expect the USEPA to be interested in the probability over time of a water quality criterion exceedance. However, the criteria actually specify recurrence frequencies of exceedances because they, unlike probabilities, can be documented and enforced.

If the probabilities of effects are subjective, the interpretation is more ambiguous. (See also Box 18.2.) You may have no data or model to estimate the relevant frequencies, but as an assessor you may provide an estimate based on your expert judgment. Because subjective probabilities are sometimes thought of as bets; you may clarify your beliefs by asking yourself what odds would you require before you would bet that a tailings dam will fail within 25 years after construction? In addition to the empirical frequency of tailings dam failures, you might subjectively consider the potential increase in large rainstorms due to climate change or your opinion of the engineering firm that designed the dam. An extreme version of subjectivism was the mining company's statement that the probability of dam failure is zero. While I have seen many probabilities described as frequencies, I have never seen a probabilistic assessment result described as the assessor's subjective degree of belief. Perhaps it is time for subjectivists to come out of the closet.

Assessors may choose to express results as probabilities because some tools for estimating frequencies are probabilistic. Monte Carlo analyses

BOX 18.2 AMBIGUOUS PERCENTAGES

Probabilities are not the only ambiguous expression of results. Percentages are also scales that have no inherent meaning and are often used ambiguously. The most familiar example is weather forecasts. What do meteorologists mean by a 30% chance of rain? Different people have told me with great certainty that it means (1) it rains on 30% of days on which a 30% forecast is issued, (2) 30% of the forecast region is expected to receive rain, and (3) the forecaster's confidence in the occurrence of rain is 30%. Also, how much precipitation is required to qualify as rain? Despite these ambiguities, percentages are generally recognized to be scaled numbers so they are not as ambiguous as probabilities in environmental assessments, which are assumed to be something more. Both require clear definitions.

require a common probability scale for parameters and Bayesian networks are networks of conditional probabilities. The formulas for calculating conditional probabilities are standard analytical tools even though people are more likely to get the right answer when a problem is presented as natural frequencies (Gigerenzer 2002; Gigerenzer and Hoffrage 1995). Frequencies are also more easily communicated, and if they are natural (not normalized), the size of the denominator indicates the size of the sampling unit and therefore is an indicator of the strength of the evidence (Chapter 25).

18.3 Combined Probabilities of Exposures and Effects

If both exposure and exposure–response relationships are expressed as distributions, the joint probability may be used as an estimate of the probability of effects. This adds complexity to the modeling and requires care to ensure that the distributions are concordant. That is, integrating them should generate a probability that corresponds to a frequency of interest. Most commonly, these are frequencies of occurrence of exposure levels and frequencies of response given exposure levels. For example, if you derive a frequency distribution of aqueous concentrations of a chemical and a distribution of acute lethality to species of fish also with respect to concentration (an SSD—Chapter 27), the joint probability can be considered the probability of a kill of a fish species with respect to concentration (Van Straalen 2002). Joint distributions are commonly used in ecological risk assessments by the pesticide industry and their contractors, and who have even developed software for them (Dreier et al. 2021).

An alternative for integration of probabilistic exposure and exposure–response relationships is to represent all of the exposure and effects processes in a single model. This bypasses calculation of joint probabilities by representing the source, exposure, and effects processes as a continuum (Chapter 5). One method is Monte Carlo simulation of a complete mathematical exposure and effects model such as AQUATOX (Park, Clough, and Wellman 2008). The other is Bayesian networks and structural equation models of causal networks that encompass exposure and effects (Chapter 23).

Both the joint probability and one model approaches were proposed in the first ecorisk methods manual (Barnthouse and Suter 1986) and have been developed since (Lombardo et al. 2015; Maund et al. 2001; Van Straalen 2002), but neither is in routine use.

18.4 Summary

Decision making under uncertainty does not require that analytical results be expressed as probabilities. Most of the assessment results that are expressed as probabilities are frequencies and are more clearly expressed in those terms. Rescaling frequencies as probabilities aids some computations, but clear communication requires expressing a real-world property.

19

Complexity

Nature is not more complicated than you think, it is more complicated than you CAN think.

—*Frank Egler (1970)*

Don't take refuge in complexity.

—*Dan Levy (2021)*

The world is indeed complicated. The ecosystems of your nation include an effective infinity of species, life stages, levels of organization, behaviors, species interactions, processes, and structures. Add the nonliving environment and you include topographies, hydrologies, contaminant mixtures, background chemistries, energy flows, material cycles, climates, seasonalities, temperatures, substrates, and other complexities. The tests and models developed for assessments are inevitably simpler than the real world. Your assessments will capture little of the complexity of real ecosystems responding to pollutants. Even if you wait for the pollutant to contaminate real ecosystems and then monitor them, you will count or measure few of the ecosystem's variable properties. Therefore, you should focus on the assessment problem and the information needs of the decision-maker, not the complexity of the situation.

When performing predictive assessments, we may attempt to emulate real-world complexity in our tests or models. For example, pond mesocosms were specified by the USEPA for tier IV testing of pesticides, because they incorporate much of the complexity of natural ecosystems (Chapter 21). However, costs were high and interpretation proved to be ambiguous. If the point of ecosystem-level testing is to test ecosystem-level endpoints like zooplankton abundance or primary production, nothing significant happened if only species composition changes. If the point is to test multiple population-level endpoints like abundance of individual species, then major effects occurred. Also, we have to ask how generalizable are mesocosms? Pond mesocosms are small for lentic ecosystems (ponds and lakes), usually lack fish, and always lack top predators. They are used to assess aquatic effects in general, but it is questionable whether they are relevant to streams and rivers or even to lakes and ponds that have fish and stratification. Each real ecosystem is complex in its own way. This is not to say that mesocosms are never useful (Chapter 21). Mesocosms are justified if an important

DOI: 10.1201/9781003156307-21

assessment problem can be emulated in a mesocosm and not in simpler systems. However, running a mesocosm test just to have complex test results is not necessarily helpful. Similarly, a complex mathematical model may raise more questions than it answers.

Ecoepidemiological and epidemiological studies for condition and causal assessments address an appropriately complex system, the real world that must be causally explained. Confronting that natural complexity often requires complex assessments, particularly when the effects have multiple possible causes. For example, in the Elk Hills San Joaquin kit fox assessment, fur samples were analyzed for contaminants, blood draws were sent for disease diagnostics, data from radio collars were used to determine whether foxes were co-occurring with contaminant sources, known sources and diffuse contamination were characterized by analysis of wastes and soils, dead foxes were necropsied for cause of death, dens were monitored for production of young, prey abundance was determined annually, and weather data were collected. Data were obtained from developed and undeveloped areas of the Elk Hills and from one developed and two undeveloped reference sites (Chapter 29). Analyses, including a demographic model, were required to turn the data into evidence for the alternative causes, and large weight-of-evidence tables were created to identify the actual cause (Suter and O'Farrell 2009). This complex multiyear assessment credibly identified predation by coyotes as the primary cause, whereas a simpler assessment had misidentified the cause as drought.

Another case of complex causal assessments is the determination of the cause of bald eagle deaths in the southeastern U.S. which took approximately 25 years and was, as with the kit foxes, justified by the importance of the species. It was found that a novel toxin in a novel cyanobacterium on an invasive aquatic plant was consumed by coots. The debilitated coots were consumed in turn by the eagles which died (Section 24.6) (Breinlinger et al. 2021; Stokstad 2021).

The even more complex climate change assessments by the Intergovernmental Panel on Climate Change are clearly justified by the important planet involved and the need for both causal explanation and prediction (see their numerous reports at https://www.ipcc.ch).

19.1 Considerations

You should address the appropriate complexity of your assessment during problem formulation. It is related to defining assessment endpoints. That is, how many endpoints are included, and can they be assessed with conventional data and analyses? It is also related to exposure. How many sources, receiving environments, and transformation processes must be assessed at what level of detail? The following considerations are relevant to decisions concerning the complexity of assessments.

Particularities of the case—Appropriate complexity can depend on the requirements of the specific case. For example, the derivation of water quality benchmarks does not normally include sensory effects. However, because of the importance of olfaction to successful migration of salmon, the sensory toxicity of copper was considered in the derivation of a NOAA benchmark (Hecht et al. 2007). Similarly, mammalian inhalation exposure to contaminated soils is not routinely considered in ecological risk assessments. However, because both the endangered black-footed ferret and its prey dwell in burrows, inhalation of vapors from contaminated soils may be considered where the ferrets may occur (Markwiese et al. 2008). More generally, if a pesticide or other agent is specifically hazardous to one taxon, any other taxa that depend on the sensitive taxon should be considered.

Practicality—You have limited time to complete an assessment, limited staff on the assessment team, and limited resources to hire employees or engage contractors. You cannot assess every endpoint, habitat characteristic, and co-occurring stressor that might seem relevant. Failure to consider practicality results in failure to complete the assessment or, if there is no deadline, results in long delays in decision making. For example, Superfund assessments are notorious for taking more than a decade to complete. Therefore, an important task of problem formulation is to manage complexity while developing a practical assessment plan.

Sufficiency—You must determine how much complexity is needed for sufficient confidence. In regulatory contexts, consensuses have developed about what effects testing and analysis is reasonable to require. For registration of new non-pesticide chemicals, the consensus has been that aquatic ecological risks are adequately addressed by tests of at least one fish, invertebrate, and algal species and application of a safety factor to the most sensitive test endpoint. To estimate exposure for new chemicals or types of sources, in most cases, you need generic simulation models of transport, transformation, and fate. However, in some aquatic risk assessments, it is sufficient to simply divide an effluent concentration by a factor to represent dilution. For non-standard assessments, sufficient complexity must be judged during the problem formulation.

Unintended consequences—More complex models or other inferences may actually make your assessment less reliable by compounding errors and uncertainties. This is particularly apparent in large simulation models (USEPA 2009). It may also be true of experimental complexity. For example, in a mesocosm test, you will see the effects of the test chemical on some species interactions, but you may be led astray when applying those results to any other ecosystem. In a review of the effects of carbaryl on interactions of larval frogs and toads with predators and competitors, I found that the results depended on the species of each and on the test design (Box 34.1 in Suter 2007). These were not just quantitative differences but actual differences in the nature of the effects. If you have only one of those tests, how would you use it in a risk or causal assessment when you do not know whether competition or predation would increase, decrease, or have no effect on larval anurans in your case? So, if you move from direct toxic effects to direct effects plus species interaction effect, you may be adding uncertainty without gaining confidence.

Parsimony—You can always appeal to Occam's razor, also known as Occam's law of parsimony. Of the several versions that are attributed to William of Occam, the most relevant to us is: "it is futile to do with more what can be done with less" (McFadden 2021). Engineers have the KISS principle, keep it simple stupid. Parsimony helps us to deal with the practicality problem by using the simplest sufficient methods. It is also argued that simple hypotheses or models are more likely to be true. A review of 97 comparisons of economic predictions from simple and complex methods in 32 papers reported that none of the papers found that complexity increased accuracy (Green and Armstrong 2015). When quantitative comparisons were made, greater complexity increased prediction errors by 27% on average.

19.2 Strategies

In general, you should judge the need for complex testing, measurement, or modeling during problem formulation by asking, could this information make a consequential improvement in the conclusion? However, certain practices help you to deal with complexity.

If assessment methods are standardized, you will not need to determine appropriate complexity. Just follow the protocol. For example, in the U.S. and most other nations water quality criteria are derived using very specific methods for species sensitivity distributions (USEPA 1985). However, when the criteria derived by standard methods are problematical, as with Se and Cu, more complex chemical-specific methods have been used (USEPA 2007a, 2016a) (Chapter 22). Those more complex methods have been resisted by some states as too complex and difficult because they require detailed site-specific water chemistry to implement a biotic ligand model (Cu) or require collection and analysis of fish tissues (Se). In other words, more complex methods may cause unacceptable costs and labor.

Environmental quality benchmarks (Chapter 22) are great assessment simplifiers. Once acceptable concentrations in air, water, soil, food, effluents, etc. are established, you need to only measure or estimate the exposure concentration by an accepted method and compare it to the quality benchmark. Then you have enough information for many assessments. The benchmark used needs to be only adequately protective as defined by legal goals. The U.S. ambient water quality criteria are stated to be protective in most places most of the time. They are like highway speed limits which are considered adequately protective even though they may not be protective in hard rain and may be overly protective for a professional driver on an open road. Benchmarks make protection possible, because routine enforcement of goals for water or air quality could not be based on derivation for each case of a case-specific benchmark.

Tests of organisms are conceptually simple. Organisms and their attributes are understood by decision-makers, stakeholders, and the public, and organism attributes for nonhuman and human animals are the same or analogous. Further, if you protect organisms, populations and communities will be protected. Things become more complex if protection is not the goal. That is, if the decision-maker wants a prediction of the nature and levels of the effects on populations or communities, you must consider how effects on organism attributes translate to population and community endpoints.

Tiered assessment is an assessment approach that was designed to minimize the use of complex and costly assessment methods (Cairns, Dickson, and Maki 1978) (Section 4.11). You begin with an assessment based on a few simple tests, basic chemical properties, and simple exposure models. If the difference between the test endpoint value and the estimated exposure level is large, in either direction, you can stop. That is, you can stop if it is clear that the chemical is unacceptably toxic or clearly nontoxic based on the margin of exposure (Section 4.4). If exposure and effects levels are not sufficiently different, more complex testing, measurement, and modeling are performed until you have sufficient confidence to draw a conclusion. Hence, in tiered assessments, you do not add complexity until you have shown that it is needed. The major limitations on tiered assessment are that you must have control over the process of generating information, and you must have enough time to go through the tiers of testing and assessment. Also, early tier tests are not necessarily predictive of chronic hazards in real-world conditions, even with safety factors. For example, the herbicide glyphosate has very low toxicity to honeybees (the standard pollinator test species) in standard early tier tests so pollinator assessment could end at a low tier. However, glyphosate disrupts thermoregulation in bumble bees which lowers brood production if food is limited, as is likely the case where herbicides are applied (Crall 2022; Weidenmüller et al. 2022). However, tiered assessment is generally a reasonable strategy.

Field studies ensure real-world complexity without the concerns about designing appropriate complexity into complex test systems and models. However, designing and conducting field studies for determining causation are complex in themselves (see the kit fox and bald eagle examples above and Chapter 21).

19.3 Conclusion

You will never create a test system or a model that is nearly as complex as nature. You will seldom be sure that increasing the complexity of your methods will result in better representation of the natural systems of interest. Further, by increasing the complexity of your methods, you will introduce opportunities for uncertainties and errors. However, in some cases, you may identify particular complexities that are essential to your assessment. Determining what is essential is difficult in predictive assessments unless the effects of the chemical or members of a class of chemicals are shown to be a result of complex dynamics. Causal assessments are less ambiguous because you keep adding complexity to your causal hypotheses until you know the cause. That process has been going on for more than a decade with respect to honeybee declines. Ultimately, the only reliable strategy is to monitor the environment to detect effects that were not predicted. The environment inherently has the correct complexity, but even then you cannot monitor everything. Determining how to appropriately handle complexity is one of the most difficult issues for environmental assessors.

20

Bias, Fraud, and Error

The truth doesn't care about our needs or wants, it doesn't care about our governments, our ideologies, our religions. It will lie in wait for all time. And this, at last, is the gift of Chernobyl. Where I once would fear the cost of truth, now I only ask: What is the cost of lies?

—*Valery Legasov, from his tapes*

If you are like most assessors, you give little thought to the influence of bias, fraud, or error in the data that you use or in your own work. Richard Feynman said that the proper attitude for a scientist is open-minded skepticism. In general, we are not skeptical enough. We tend to accept anything published in peer-reviewed literature and are only a little more skeptical of findings in unreviewed reports. We are also disinclined to question our own biases and reluctant to take the time to quality assure our work.

20.1 Concepts

Bias favors one result over another and may be explicit or implicit. It is explicit if you know the desired result and make assumptions, choose data, or by other means make things come out "right," because you believe that you know what "right" is. Bias is implicit if you do not have conscious intent. Biases may come from your employment, experience, ideological beliefs, or simply a desire to get a publishable result.

Fraud may be an extreme result of bias or of being too lazy, incompetent, or rushed to do the work right. Fraud may include inventing or deleting data or model results, misrepresenting the conclusions of a cited study, deliberately excluding a relevant study, or other forms of cheating.

Error is the result of making mistakes or not working carefully. For example, we have all made decimal point errors or typed the wrong cell range into a spreadsheet equation. We can only hope that we or a reviewer catch our errors before publication. Errors are common in the scientific literature. For example, a toxicological study in a peer-reviewed journal reported 200% survival in the controls. When I contacted the author, he said that the same error appears in his doctoral dissertation, so he and his committee members missed it, and then it slipped by the journal's peer reviewers. Cognitive

DOI: 10.1201/9781003156307-22

errors are more subtle than these cases of slip-ups. They are errors in thought that occur when we use the answer that seems right rather than one that comes from analytical thinking. This is Kahnman's (2011) cognitive system 1 versus system 2 distinction. Formal analytical methods are intended to minimize cognitive errors by keeping you in system 2 (Norton, Rao et al. 2003). [Note, this is the common English usage of error as the result of mistakes, not the statistical usage—the difference between an obtained value and the true value.]

Political Influence: You might say that political influence is just another bias. Political bias is apparent in cases that I witnessed such as the USEPA's changing positions on chlorpyrifos and the Pebble Mine between the Obama, Trump, and Biden administrations. However, it goes deeper than that. The laws and regulations may favor industry or health and the environment depending on who has the power when they are written. The Clean Water Act strongly protects the physical, chemical, and biological integrity of the nation's waters, but the clean air act protects only human health and welfare. The difference reflects the politics at the time the laws were passed. The Toxic Substances Control Act (TSCA) was written at the direction of the chemical industry. It was so biased that it has been impossible to ban non-pesticide chemicals in the U.S. and even research on chemicals in the environment has been impeded (Wagner and Gold 2022). The "production of ignorance through regulatory structure" is described with respect to per- and polyfluoroalkyl substances (PFASs) by Richter et al. (2021). The 2016 Lautenberg Act amendments have somewhat strengthened TSCA, but the record is still not encouraging. The biases in laws and regulations are beyond your control as an assessor. However, that is not a reason to give up. Chlorpyrifos was ultimately banned for most uses (see below), and even Trump turned against the Pebble Mine in the end because his son and some other Republicans like to fish in the Bristol Bay watershed. Political influence may have surprising motivations.

20.2 Examples

We would like to think that we do unbiased science and that our colleagues are trustworthy, even if they are on the other side. Most of the publicized examples of scientific fraud involved biomedical or psychological research, but the environmental sciences are not exempt. A few examples may help you to develop a healthy skepticism.

A paper in *Science* claimed to show that larval fish preferred microplastics to natural prey, resulting in decreased growth, lethargy, and failure to avoid predators (Lonnstedt and Eklov 2016). The authors were accused by their colleagues of not having preformed the described studies and they could not produce the original data (Enserink 2017). The authors attributed the actions of whistle blowers to jealousy. After an investigation by the Central Ethical Review Board in Sweden, the paper was retracted (Enserink 2017).

Another paper in *Science* that was part of a study of the estrogenic effects of contaminants on alligators and turtles in Lake Apopka, Florida, found 1,000-fold synergism between hydroxylated polychlorinated biphenyls and multiple chlorinated pesticides in the yeast estrogen assay (Arnold, Klotz et al. 1996). However, neither independent investigators nor the original investigators could replicate that result, so the paper was withdrawn (McLachlan 1997).

A study claimed to show that Florida panthers, an endangered subspecies, prefer large patches of forest habitat and avoided passing through non-forest (Maehr and Cox 1995). A science review team found that the result was obtained by removing data that were not consistent with the hypothesis. However, the journal's editor refused to retract the paper, the author continued to serve as a contractor to developers, and regulators continued to issue exemptions to the developers based on the paper (Pittman 2020).

Environmental Science and Technology retracted two papers that exaggerated air pollution due to hydraulic fracturing (fracking) and due to the Deepwater Horizon oil spill (Paulik, Donald et al. 2015, Tidwell, Allan et al. 2015, Chawla 2016). The authors claimed honest calculation errors in both cases. Both papers were republished with changed results under other titles (Paulik, Donald et al. 2016, Tidwell, Allan et al. 2016).

A group of authors spent three years attempting and failing to replicate papers by another group of authors that reported behavioral effects of ocean acidification on recovery of corals and fish (Enserink 2020). As in many other failed replication studies, the original study authors attributed the failures to small differences in methods. Even if that is the explanation, it is problematical. Results that can be generalized to the real world cannot be specific to small differences in container volume, duration, or other details of a study. Since then, more studies have resulted in accusations by analysts and colleagues that the original study authors engaged in data fabrication (Enserink 2021). Now a university investigative panel has also reported fabrication and falsification, and a paper in Science has been retracted (Enserink 2022).

An assessment of the benefits of marine protected areas has been retracted and has brought down the most prominent figure in U.S. marine ecology (Mossler 2021). It concluded that closing an additional 5% of the ocean to fishing would increase fish catches by 20% (Cabral, Bradley et al. 2020). That conclusion has been shown to be based on errors and impossible assumptions. It was published in the Proceedings of the National Academy of Sciences (PNAS) with Jane Lubchenco, former Director of the National Oceanic and Atmospheric Administration (NOAA) and White House Deputy Director for Climate and Environment, editing the paper including choosing the reviewers. One of the authors was her son-in-law and former graduate student. Lubchenco was also collaborating with authors of the retracted paper on a companion paper that built on its analyses. She has now been sanctioned by the National Academy of Science. The companion paper (Sala, Mayorga et al.

2021) is said to have similar errors and assumptions but does not have the same peer-review scandal and has not been retracted.

The USEPA's no-observed-adverse-effect-level (NOAEL) for chlorpyrifos was taken from a 1972 deliberate human dosing study funded by Dow Chemical. A recent review of that study found that the statistical analysis, performed by a Dow statistician, contained errors, including incomplete data for the NOAEL level (Sheppard et. al. 2020). Further, Dow had convinced the USEPA to not peer review their unreviewed and unpublished study. For over 15 years, the USEPA failed to notice the data exclusion issue and failed to reanalyze the data to correct for differences in duration of different dose levels. A 2016 reassessment by the USEPA finally determined that chlorpyrifos should be banned for food crops. That finding was blocked by the first Trump-appointed USEPA administrator and reinstated by the Biden-appointed administrator.

The conflict over ecological risks from atrazine, particularly deformities in frogs, has been well documented and unusually nasty (Suter and Cormier 2014, Suter and Cormier 2015a). A review of animal studies found that there was no difference in risk of methodological bias (that is, the reported methods were equally reliable), but non-industry-supported studies were much more likely to report statistically significant effects (Bero, Anglemyer et al. 2016). Such funding bias should prompt careful independent analyses of the causes of differences in results.

I devoted so much space to these examples because colleagues have told me that bad science is common in other fields, but not in environmental science and assessment. Not true, so you need to be alert. These are just a few conspicuous recent cases.

20.3 What to Do About Information Sources

You, as an assessor, should consider the following practices to minimize your use of biased, fraudulent, or erroneous information (Suter and Cormier 2014). Some of these recommendations are time consuming and can be limited to those studies that provide highly influential information.

- Do not rely on peer review—assessors must consider peer review to be a contributing factor for rating data quality, but they should still treat every publication with skepticism.
- Perform and document your own careful review of each influential study.
- For studies that have not been peer reviewed or that are controversial or influential, consider engaging a peer-review panel to focus on the suitability of the study for use in the assessment. This should supplement, not replace, your own review.

- Check for retractions, rebuttals, and corrections (search using CrossRef digital object identifiers [DOIs], the Retraction Watch database, or any of the reference managers that flag publications in the Retraction Watch database). Do not use retracted or uncorrected flagged data. Also, carefully assess the validity of letters to the editor criticizing publications and responses to those criticisms.
- Check declared sources of funding and conflicts of interest and search the web for evidence of other potential funding or personal biases or undisclosed conflicts of interest. At least be alerted to the need for rigorous review when funding sources are not neutral. When the volume of literature permits, compare results from studies funded by different sources for signs of bias and report the results.
- Require full disclosure of methods, in the publication, in supplementary materials, or provided on request. Because most journals have page limits, the methods sections of papers are summaries.
- Be skeptical of studies that are not replicated or at least consistent with similar studies (e.g., with other species, durations, routes of exposure).
- Extract data from primary, not secondary, sources—reviews and other secondary sources can introduce error as well as potential bias.
- Obtain original data and reanalyze them for critical information.
- Focus on estimation and confidence in estimates, not hypothesis testing, which is easily manipulated and misleading.
- Use your review of the data source to assign a weight to each piece of evidence derived from the data and incorporate the risk of bias into any weight-of-evidence analysis. Neither peer review nor statistical hypothesis testing answers the question, how worthy of belief is the evidence?

20.4 What to Do About Your Own Biases and Errors

The following long list of suggested practices should help you control your biases and minimize errors (Suter and Cormier 2015).

20.4.1 Objectivity

The following suggested practices are intended to ensure that assessors work in a disinterested manner so that the results are objective.

- Check for personal predilections (be self-aware) and suppress them (be self-skeptical).
- Put yourself in the place of an assessor from another sector (industry, health or environment advocacy, or government). What would they do, and might they be right?
- Consider the opposite. For each decision in the assessment, stop and consider the arguments for at least one strong alternative assumption or method.
- Use standard frameworks and methods for performing assessments unless justified by the circumstances. Ad hoc methods are an opportunity for bias.
- Consult the decision-maker during the planning and scoping phase to establish aspects of the scope of the assessment that might bias the results if determined by assessors.
- After the planning stage, minimize contact with the decision-makers to avoid management bias. However, meetings for clarification may be needed, particularly if decision-makers change or new issues arise.
- Work in a group. Group members error-check each other, combine knowledge and skills, and increase the range of options considered, thereby reducing the bias of an individual. However, groupthink can also be a source of bias, so also think things through alone.
- Systematically search the literature for relevant data and information. Basing an assessment on the evidence that comes readily to mind or is easily accessed can bias the results.
- Objectively screen and weight the available information.
- Use consensus science unless strong justification supports marginal science. The consensus of the scientific community is not always right, but it is generally the best judge of good science.
- Perform and report sensitivity analyses for questionable assumptions and data sources. That is, if an assumption is controversial or a data set is from a potentially biased source, determine how much difference it makes in the assessment's results.
- Avoid interpretation of results in terms of values unless the value judgments are defined by law or existing policies or specified in the planning process.
- Do not treat statistical test results as definitive of significance. Statistical significance is often misleading and should never be confused with biological, public health, or ecological significance.
- Weigh the evidence in support of credible alternative conclusions.
- Perform peer reviews of assessments.

- Perform separate reviews of the code for models and data analyses to determine that it performed the calculations described in the methods section.
- Replicate data extraction, influential modeling, and statistical analyses to catch errors.
- Respond appropriately to legitimate scientific challenges and do not dismiss them.

20.4.2 Technical Transparency

Transparency is the practice of clearly explaining all methods and of making available not only the assessment but also the inputs to the assessment. Its primary function is to reveal to the readers, including reviewers and critics, the potential sources of bias or error. This discipline can also make assessors more self-reflective and less likely to bias their work.

- Document and justify assumptions.
- Document and justify data generation and analytical methods.
- Document and justify inferential methods including criteria for acceptance of evidence, methods for weighing evidence, and so on.
- Document and justify departures from standard assessment frameworks.
- Make data and models available.
- Make external peer-review comments and the disposition of the comments publicly available.

20.4.3 Ethical Transparency

Ethical transparency is the practice of revealing any financial or personal circumstances that might incline you to bias the assessment. This transparency is a particular issue for assessments that are published in journals rather than as reports of the organization for which it was performed.

- Disclose potential conflicts of interest. A conflict of interest exists when ethical or legal obligations to objectively assess an issue conflict with financial interests or personal commitment to a cause.
- Report the impetus for undertaking an assessment including funding sources. Whether or not an assessor believes that the assessment results are biased by the employer or client who requested or funded it, the impetus for performing an assessment should be fully disclosed in the assessment document. This is particularly important

for assessments published in journals as reviews, editorials, critiques, or commentaries.
- Document interactions between groups such as government assessors, decision-makers, stakeholders, consultants, applicants, and responsible parties. This documentation need not be included in the assessment but should be available.

20.5 Policies and Procedures

Many governmental and industry organizations have policies and procedures to assure quality and integrity. For example, the USEPA has a large data quality program (https://www.epa.gov/quality) which is intended to assure the quality of information generated and used by the Agency. Data quality has been controversial in recent years because industry-backed policies have prohibited the use by federal government assessors of information for which the original data are not available. However, proprietary industry data were exempted. Data availability is desirable, but some important epidemiological studies do not release original data due to privacy agreements with the participants.

The USEPA Scientific Integrity Office (2012) provides guidance to Agency scientists for assuring integrity. Guidance for large organizations like this tends to be somewhat vague and general or focused on paperwork and procedures. Look for specific guidance relevant to your activities such as those for chemical analyses or peer reviewing. You should consider developing procedures that are important to your assessments and were missed by developers of general guidance. For example, neither data quality guidance nor systematic review methods in the USEPA require checking cited publications for retractions, corrections, or expressions of concern. Hence, you should go beyond policies and procedures to avoid using biased or erroneous results.

20.6 Integrity and Credibility

You can get reliable and defensible results only if you minimize errors, suppress bias, and never commit fraud. However, in this time of distrust of authority and expertise, that is not sufficient to gain credibility and trust in those results (Mebane, Sumpter et al. 2019, Brock, Elliott et al. 2021). The transparency advocated above serves to inhibit bias and fraud and allow other parties to check results and catch errors. To gain trust, transparency should be extended to all stakeholders and the public. In addition, people must feel that they have been included and treated fairly. In the Bristol Bay watershed assessment, this included collaborative problem formulation (Chapter 4), and multiple listening visits to native villages, coastal fishing towns, and the city of Anchorage by assessment scientists, USEPA managers, and two USEPA Administrators.

An advantage of being an assessor is having influence in environmental decision making (Chapter 2). However, it involves you in politics, both organizational politics and literal governmental politics. During my career, the G.W. Bush, Obama, and Trump administrations all pressured USEPA employees to minimize regulation of polluters and overrode the findings of some science-based assessments. Even pro-environment administrations may respond to industry or public pressures, such as pressures to maintain coal industry jobs. Obama put a friend from the University of Chicago, Cas Sunstein, in charge of reviewing regulations despite his ideological anti-regulation position. Industry and consulting jobs also pose political challenges to your integrity. I know that contractors often feel management pressure to please their clients. In general, the best strategy is to do your assessment work with integrity and accept that the decision-makers respond to the political power structure. If the distortion of the science is sufficiently blatant, you can resist by going to the ethics officers in your organization or even take the nuclear option and go to the press, to environmental advocacy groups, or to organizations like Public Employees for Environmental Responsibility.

Part III

Methods for Environmental Assessment

Chapters 21–31 deal with methods and methodological approaches to particular aspects of environmental assessments. The point of this part is not to get into the technical weeds. When the time comes to use Monte Carlo simulation, for example, there are computational tools and even YouTube videos. Rather, I have tried to describe the types of methods that solve particular problems.

DOI: 10.1201/9781003156307-23

21

Sources of Information

Frameworks, concepts, and inferential methods are, at least for me, the most interesting parts of the assessment toolbox. However, they all depend on information from the literature, measurements, observations, or expert judgment. Information is data or other facts used to derive evidence.

21.1 Existing Information and Literature Reviews

Your primary source of information for an assessment is likely to be the existing literature. Traditionally (i.e., when I was starting out), reviews began by consulting experts in the topic of interest. Often you were your own expert and you found your own way into the literature. This would lead to assembling some papers, and the references cited in those publications would lead to other relevant publications and perhaps to helpful experts. Such reviews have no systematic method. More recently, searches of bibliographic databases have become dominant. This reduced the effort required to assemble a large body of literature but created a problem of reviewing and screening the hundreds or thousands of hits. In addition, electronic searches by different groups still produced reviews with different results due to differences in databases, incomplete searches and inclusion criteria, and inclusion of inappropriate studies. This is a particular problem for assessors because the quality of our reviews may determine a chemical's benchmark value or hazard designation. Your choice of studies to include in a review may be biased or you may be accused of biasing a review if you were not rigorous and transparent. See for example, the case of atrazine in which differences in study inclusion were attributed to bias (Rohr and McCoy 2010). Similarly, systematic reviews of the genotoxicity of glyphosate by the USEPA and IARC came to opposite conclusions. A review of the reviews found that the agencies included different studies (Benbrook 2019). In particular, the USEPA relied heavily on unpublished, registrant-commissioned standard studies, while IARC emphasized journal-published papers with nonstandard methods. In other words, some biases are based on policies concerning what sources are most reliable.

One solution is to skip the literature and use an existing database such as the USEPA's ECOTOX or Health and Environmental Sciences Institute's

DOI: 10.1201/9781003156307-24

EnviroTox which contain extracted test endpoint values for aquatic and terrestrial organisms. If you take this approach, you are accepting the completeness and quality of the database. Because they are created by trained literature reviewers following a standard method, they are reasonably reliable, but their primary advantage is the efficiency of relying on a standard literature review system. Another type of database is data repositories such as Gene Expression Omnibus (https://www.ncbi.nlm.nih.gov/geo/). Repositories depend on the compliance of data providers with rules and criteria. A similar solution is to use recently published reviews of individual chemicals or chemical classes performed by agencies, manufacturers, industry groups, or academic scientists. These reviews may not have used standard or formal methods and they may be old, so their methods and recency must be considered.

A solution to the problems with literature searching is to use a structured and transparent literature review processes that produces systematic reviews. Systematic reviews should ensure that (1) nearly all relevant studies are retrieved by a search, (2) exclusions from the search results occur for clear and consistent reasons (e.g., not freshwater, not peer reviewed, or not in English), and (3) relevant studies that are unreliable are screened out (e.g., original data are not available).

One limitation of literature reviews is that they find papers in journals but often miss reports from government agencies and other organizations. It is worthwhile to separately search the websites of relevant organizations.

Systematic reviews were originated by the Cochrane Collaboration in an attempt to assure complete and high-quality literature reviews for meta-analyses of clinical trials and to minimize discrepancies among reviews of the same medical treatment (Higgins et al. 2021). Meta-analyses are statistical analyses that combine multiple studies, usually to generate an aggregate mean and confidence interval (Chapter 24). This original systematic review method is relatively straight-forward because it includes only clinical trials of a treatment for a particular medical condition. In contrast, environmental assessments involve various taxa, levels of organization, environmental conditions, forms of the agent, and methods for generating the information. Even environmental toxicological reviews for human health may include both animal and human data generated under various conditions. As a result, developing criteria for inclusion of studies can be complicated.

Systematic literature reviews have three fundamental steps:

1. Literature searches depend primarily on computerized searches of literature databases such as PubMed, Google Scholar, and Web of Science. Your success in such searches depends on carefully designing the search terms and the Boolean operations (and, or, not and, not or) that define associations between terms. You may also verify

the search by checking for studies that you know should be included or by searching reference lists for more studies that the computer search should have found.

2. Data screening determines whether the information in the studies is relevant to the assessment and whether the studies include such serious flaws that they are unusable. It may also assign the information to useful categories such as animal versus human for human health assessments, terrestrial versus aquatic for ecological assessments, and route of exposure for either.
3. Data extraction is the process of identifying the relevant data and qualitative information and extracting them to templates. This is an opportunity to identify errors in the data. When extracting ecotoxicity data for screening chemicals at contaminated sites, I have found impossible values in the peer-reviewed literature such as 75% survival in a treatment with ten animals. Ambiguities, errors, and missing data may lead to rejection of a study or to correction by contacting the authors.

The characteristic that distinguishes systematic reviews from prior methods for assembling existing information is that the method is defined and reported. The result of a systematic review should be a data set that is nearly complete, unbiased, and reproducible.

After these three steps, use the collected data or qualitative information to perform an inference. The inferential method depends on the endpoint and the information that you manage to collect (Chapter 24). Some systems go beyond the systematic literature review and include the inferential step as inherently part of the systematic review. For example, the USEPA's (2018a) systematic review guidance for non-pesticide chemicals includes two steps beyond the three listed above, data evaluation and data integration and summary, that are effectively weight of evidence.

Systematic reviews are a major component of chemical assessments by the USEPA, EFSA, ECHA, and many other regulators. Agencies and even individual offices within agencies have their own methods such as USEPA (2018a). The development of software to support systematic reviews has become a significant enterprise. The software used by the USEPA for systematic reviews of chemical hazards currently includes HERO for literature searching and storage; SWIFT for study sorting and prioritization based on machine learning; and HAWC for tracking results of multiple reviewers, data storage, and presentation of results. Artificial intelligence promises to automate much of the systematic review process. This is important because systematic reviews of well-studied chemicals or treatments can be extremely time consuming. I know of two projects that were never completed because the effort required by the systematic review used the entire budget and allotted time.

21.2 A Case Study of Literature Review for Meta-Analysis

Although meta-analyses based on systematic literature reviews are intended to resolve scientific issues, they have been remarkably contentious. The researchers have an interest in whether the reviewer's results support their study's conclusions.

An example is provided by a meta-analysis of studies of trends in insect abundance, a commentary on that meta-analysis, and a response to that commentary (vanKlink, et al. 2020a, b). The meta-analysis found that terrestrial insects were declining at ~9% per decade and aquatic insects were increasing at ~11% per decade. Commenters rejected the increase in aquatic insects and raised issues with the systematic review. They pointed out that several natural and designed experiments were included which are indicative of effects of a particular treatment rather than of global trends in abundance. For example, one included study monitored insect abundance in response to a dragonfly habitat restoration project. The commenters also objected to studies in exceptional circumstances such as the area around Chernobyl or the creation of a polder or reservoir. Although the meta-analysis nominally addressed insects, 27 of 63 freshwater communities included macroinvertebrates in general. The commenters concluded that ecologists must apply more rigorous standards for systematic review. The review's authors responded that they had established rather broad criteria for study inclusion and stuck with those criteria in order to avoid "cherry-picking" and "mission creep." Adding new criteria after seeing the data would be equivalent to post hoc statistical analysis, which is cheating. Besides, they state that the commenters' issues made little difference in the results. Terrestrial insect numbers still went down, and aquatic insects went up when they performed sensitivity analyses of the commenter's suggestions. That is good news for those of us who have worked to improve water quality; our efforts are working.

Although that review was problematical in many ways, it was an improvement over a prior systematic review of insect trends with many simple errors and one obvious and major error (Sánchez-Bayoa and Wyckhuysb 2019). Those authors used the search terms "insect," "survey," and "decline" but did not search on "increase." Not surprisingly, they concluded that declines in insect abundance were severe and ubiquitous. They were widely criticized including one pointed critique titled: *Alarmist by bad design: strongly popularized unsubstantiated claims undermine credibility of conservation science* (Komonen, Halme, and Kotiaho 2019).

These are the sort of errors and biases that systematic review was developed to prevent. Literature reviews are essential, and with a meta-analysis or qualitative inference, they can constitute an environmental assessment. However, you must be willing to put in the time and effort to get it right.

21.3 Laboratory Tests and Measurements

Assessment-specific determinations of chemical properties or toxicities are most often obtained from conventional laboratory studies. These include single species toxicity tests and measurements of chemical properties such as solubility. These studies may also address simple mixtures such as pesticide formulations. Results of these studies are likely to be accepted if a standard method is used. The USEPA and some other agencies have their own standard methods, and the Organization for Economic Cooperation and Development (OECD) has numerous international consensus methods (OECD 2022c). Nonstandard methods may be used if no standard method is appropriate, but considerable effort may be required for method development and validation. That effort is justified if, for example, you have observed a distinct toxic effect in the field that is not a standard test endpoint, and you wish to determine whether a suspected causal agent can induce that effect.

The principal advantage of laboratory studies is control of the conditions and treatments. In toxicity tests, the random assignment of organisms to defined treatment levels including controls provides assurance of causation. Another advantage is the consistency and comparability of laboratory results across chemicals and studies. Also, the costs of typical laboratory studies are usually low relative to field studies of toxic effects.

In laboratory tests, you should carefully consider the relationship between the exposure in the laboratory and the field conditions being assessed. For example, the OECD provides two insect toxicity tests of aqueous sediment (OECD 2004a, b). One spikes the water and the other spikes the sediment. Spiked sediment might best represent sediment from pesticide-contaminated soil eroded into a stream or pond. Spiked water might best represent an aqueous effluent of a pesticide production plant. Other source scenarios require similar consideration. Ideally, the flexibility of the test results would be optimized by reporting multiple exposure media such as surface water, pore water, and solid phase in the OECD sediment in water test.

Laboratory toxicity tests for human health risk assessment are largely limited to rat or mouse tests. However, other animal species are also used as appropriate. For example, fish embryos are used in teratogenicity tests.

The USEPA and other agencies and organizations are exploring a variety of in vitro tests and computational techniques collectively referred to a New Approach Methods (NAMs). These methods are developing rapidly and becoming important sources of information. However, they have not yet been integrated into mainstream environmental assessment. Therefore, they are briefly discussed in Chapter 31, The Future.

21.4 Model Ecosystems

The principal disadvantage of conventional laboratory studies is their simplicity relative to nature or to human variability (Chapter 19). A potential solution to this problem in aquatic ecotoxicology is tests performed in model ecosystems: microcosms and mesocosms. Microcosms are laboratory systems that include multiple species and perhaps multiple media and some habitat structure. A typical microcosm is a small aquarium or beaker with pond water and sediment and the organisms that came with them. Mesocosms are outdoor systems that include multiple media and species. Typical mesocosms include artificial ponds with natural pond sediment and biota and small channels/streams with gravel, flowing water, and benthic invertebrates. At least in the U.S., microcosms are less advocated and used than mesocosms to support assessment and regulation of chemicals.

Despite all of the work that has gone into developing and standardizing these systems, their regulatory use has been uncommon. The justification for model ecosystems is that we want to protect ecosystems so we should perform tests in ecosystems. But they raise the question, are they representative of the ecosystems that we wish to protect from toxic effects? Given the different depths, nutrient levels, suspended sediments, species, trophic compositions, etc., in an aquarium or a 5 × 5m square pond versus a lake, what can we assessors conclude? Model ecosystems have also raised questions about appropriate endpoint effects. If zooplankton abundance is unchanged but cladocerans decrease while rotifers increase, is that adverse? If algal species composition or primary productivity changes, is that adverse? Does it matter whether the changes in algal species are caused by direct toxicity or by changes in zooplankton grazing? Does it matter whether the change in primary production occurs in an oligotrophic, mesotrophic, or eutrophic water body? Such questions and objections to cost caused the USEPA to drop the standard mesocosm requirement for Tier IV testing of pesticides (Tuart 1988; Tuart and Maciorowski 1997). However, model ecosystems still have their advocates who have addressed these issues (Giddings et al. 2002), and they are still sometimes used, particularly in testing pesticides.

Mesocosms have been useful for certain well-defined problems. For example, most of the assessment endpoints for effects of herbicides on phytoplankton have been measures of productivity. Because of the diversity of phytoplankton, mesocosms have been useful for assessing that endpoint (Giddings et al. 2005). Mesocosms are also an important research tool in environmental toxicology. Clements and his many students have focused on developing a deep understanding of how metals affect aquatic insects in the mining-contaminated streams of the metal belt in Colorado. They have used various tools, but their realistic stream mesocosms have proved essential in understanding the sensitivity of stream assemblages relative to laboratory tests (Clements et al. 2019). Similarly, the criticism that generic mesocosms do not represent any actual ecosystem can be avoided by designing mesocosms to emulate a particular exposed ecosystem type. For example, mesocosms have been designed to physically emulate pesticide spray drift to the ditches that drain Dutch potato fields (Arts et al. 2006).

In the end, model ecosystems should be used when they are suited for purpose. For example, manufacturers may believe that the benchmarks derived from standard laboratory tests are over-protective. One response to that belief is to perform a "more realistic" mesocosm test that may include compensatory processes, sequestration, and enhanced biodegradation. Also, if the mechanisms causing effects in the field are complex and ambiguous, mesocosm tests may elucidate those mechanisms. Finally, in Europe, mesocosm data may be used, under certain constraints, to derive water quality benchmarks (EC 2011).

21.5 Tests of Effluents and Ambient Media

Assessing the toxicity of aqueous effluents is complex because the effluents themselves are complex mixtures. The usual way to deal with this complexity is to determine the concentrations of each potentially toxic chemical, determine the degree of toxicity, and then combine them using a simple model (Chapter 29). However, if you have the option, testing the whole effluent should give a better result, particularly if you do not have good toxicity data for all effluent constituents (Chapter 10). Realism is increased if you use the receiving water as the diluent of the effluent to create the exposure–response series.

Assessing the risks from contaminated ambient media presents a similar problem to assessing effluents. A contaminated soil, sediment, or water can be more reliably assessed by testing it as a whole than by using toxicity data for individual chemicals. The tests developed for aqueous effluent testing are appropriate for testing ambient waters. Standard tests for water, soil, and sediment from contaminated sites have been available for decades (USEPA 1994) and are now numerous (OECD 2022c). Tests for dietary exposure of wildlife on contaminated sites are rare but not unknown. For example, fish from PCB- and Hg-contaminated waters draining the Oak Reservation were fed to gravid mink, and lead-contaminated sediment from the Coeur d'Alene river was fed to waterfowl (Halbrook, Brewer, and Buehler 1999; Beyer et al. 2000).

21.6 Field Experiments

You can obtain realistic information concerning transport, fate, and toxicity of chemicals or other agents by performing field experiments. There are two general approaches. First, you can treat areas of the ecosystem type of concern with controlled levels of the agent. This approach has been primarily used to support assessments of pesticides by simulating prescribed applications (USEPA/PMRA/CDPR 2014). For example, the pollinator assessment to support the registration review of imidacloprid included exposures of honey bee hives in the field by supplemental feeding with spiked sugar water, exposure in "tunnels," mesh tents which limit foraging to treated crops, and exposure of unconfined hives to treated crops (Housenger et al. 2016).

The second approach is to expose organisms to already contaminated media. Plants are immobile and therefore can be planted in soils with various types or levels of contamination to determine effects on survival, growth, and uptake of the contaminants. Similarly, mussels and clams can be readily caged in selected locations for toxicity and uptake determination. Nest boxes may be set out for birds to determine reproductive success and contamination of the young or the food brought to them. Stream invertebrates may be exposed in wire mesh gravel boxes. These methods are not standardized, but have long been recommended (USEPA 1989). Therefore, you can be creative in developing appropriate experiments for your case, but in the field, there are always complications that must be worked out. For example, at Oak Ridge, we found that rat snakes were amazingly good at climbing posts to consume starling nestlings in nest boxes.

Field experiments provide some degree of control of exposure in realistic conditions. However, when exposing organisms to field-contaminated media, randomization is not possible and replication may be difficult. For example, if you are placing caged fish or invertebrate substrates on a stream gradient below an outfall, the dilution gradient is confounded by any habitat gradient.

Field experiments are not performed on humans for obvious reasons. However, the potential relevance of results of ecological field experiments should be considered in human health assessments. They can provide information on exposure, such as persistence, partitioning, and bioaccumulation, and possibly even toxicity if vertebrates are included.

21.7 Field Observations

Field observational studies provide the most realistic and least controlled information for your environmental assessment. The use of such studies to identify effects and their causes is known as epidemiology and ecoepidemiology when applied to humans and nonhuman organisms, respectively. (Terminology can get confusing when epidemiologists refer to studies of humans as ecoepidemiology if the studies take a broad view encompassing human populations and their habitats (March and Susser 2006).) The obvious disadvantages of such studies are the lack of control and often lack of replication. They lack control because you cannot decide which organisms, populations, or communities will be exposed, how they will be exposed, and to what degree they will be exposed. They lack replication because, in most cases, there is only one source that is being studied. If you want to make an inference about the effects of exposure to sewage effluents in general, you must find true replicate sewage outfalls for your observational study. However, there are many instances in which you are interested in only one source. For example, if you are assessing the effects of mercury contamination from a former mercury-cell chloralkali plant on consumers of fish, you can sample fish below and well above the plant. The mean and distribution of mercury concentrations in individual fish can be used to estimate risks to fishermen or to piscivorous wildlife at that plant.

The major weakness of observational studies is uncertain causation (Chapter 12). Because exposure is not controlled or randomized, it may be confounded by other factors. For example, I engaged a contractor to perform a study of metal bioaccumulation by small mammals on a contaminated site. They performed collections at contaminated locations in the spring and reference locations in late summer. Seasonal differences in soil and plant conditions and ages of the animals could confound the location differences, particularly in a semi-arid Mediterranean climate. Because it was not possible to control for confounders, I completed the assessment without that information. Other cases of confounding are less obvious, and it is important when using observational data that you identify potential confounders and control for them if possible.

One alternative to trying to demonstrate causation in observational studies is to use them as preliminary screens. That is, perform a synoptic survey of all potential causes to determine which are associated with the effects. Then perform focused studies of those that are spatially associated. The focused studies should be experimental, but they may be field studies that focused on generating evidence of exposure, sufficiency, and characteristic effects such as molecular biomarkers. With evidence of those characteristics of causation (Chapter 12), you can reliably assign causation if you have included the cause in your synoptic survey.

Designs of human observational studies (i.e., epidemiology) are relatively few and standardized. The major types are:

cohort—identify exposed and unexposed individuals and follow them over time measuring effects,

case control—identify individuals with and without existing effects and retrospectively estimate exposure,

cross-sectional—measure existing effects and current exposure levels at the same time, and

ecological—compare groups based on exposure, rather than individuals, and measure effects.

These types are described in any epidemiology text or website, but an environmental risk focused reference is USEPA (2016d). Ecological study designs are considered the least useful type for human health, but they are conventional for ecological assessments. The basic design is comparison of exposed populations or communities to unexposed or minimally exposed reference. However, in some cases, ecological receptors are treated as individuals, so the other epidemiological designs may be appropriate. For example, in the Elk Hills San Joaquin kit fox study, the foxes were fitted with tracking collars so their individual movements, diets, fecundity, causes of death, and elemental exposures were determined (Suter and O'Farrell 2009). The study design was cross-sectional. The data from individual foxes were valuable for determining causes, but a demographic model was required to estimate population trends.

Another way to step up from ecological designs is sampling along a gradient of exposure so as to derive an exposure–response relationship. In streams, you can combine a gradient downstream with an upstream reference. Gradients are usually spatial, but if exposure is changing over time as is the case when assessing recovery following a remedial action, temporal gradients may be studied alone or in combination with spatial gradients. If a study begins before the action to be assessed, you can sample upstream and downstream of the planned effluent release point and then resample after the project is completed and the effluent is being released. This Before-After Control-Impact (BACI) design corrects the upstream–downstream differences for inherent differences in the habitat quality.

Designing and conducting ecological field studies require knowledge of basic natural history. For example, studies of stream invertebrates must consider that many mayflies and some other taxa are annual. They emerge in the spring and occur as eggs and tiny early instars in the summer which do not appear in samples. Therefore, summer samples underestimate species richness and should not be compared to samples from other seasons. Animals with large home ranges do not lend themselves to ecoepidemiological studies unless, as in the kit foxes studies described above, the home ranges can be determined and related to the distribution of contaminant sources.

Gathering information from the field is not always that difficult, and it need not involve expensive studies, measurements, or even quantification. You can visit an impaired or contaminated site and look for sources and determine the spatial relationship between sources and impaired biota. In site-specific studies, I have seen seepage from the toes of landfills and slicks on streams and even smelled sources of organic contamination. I have observed physical disturbances that could confound effects of contaminants. I have observed multiple individuals of an endangered species at a proposed project site, which was sufficient to cause a project redesign.

A particularly good example of the power of qualitative field observation is provided by the causal assessment of the Willimantic River, Connecticut (Bellucci, Hoffman, and Cormier 2010). A screening assessment by a one-day workshop identified a likely location of the contaminant sources based on the

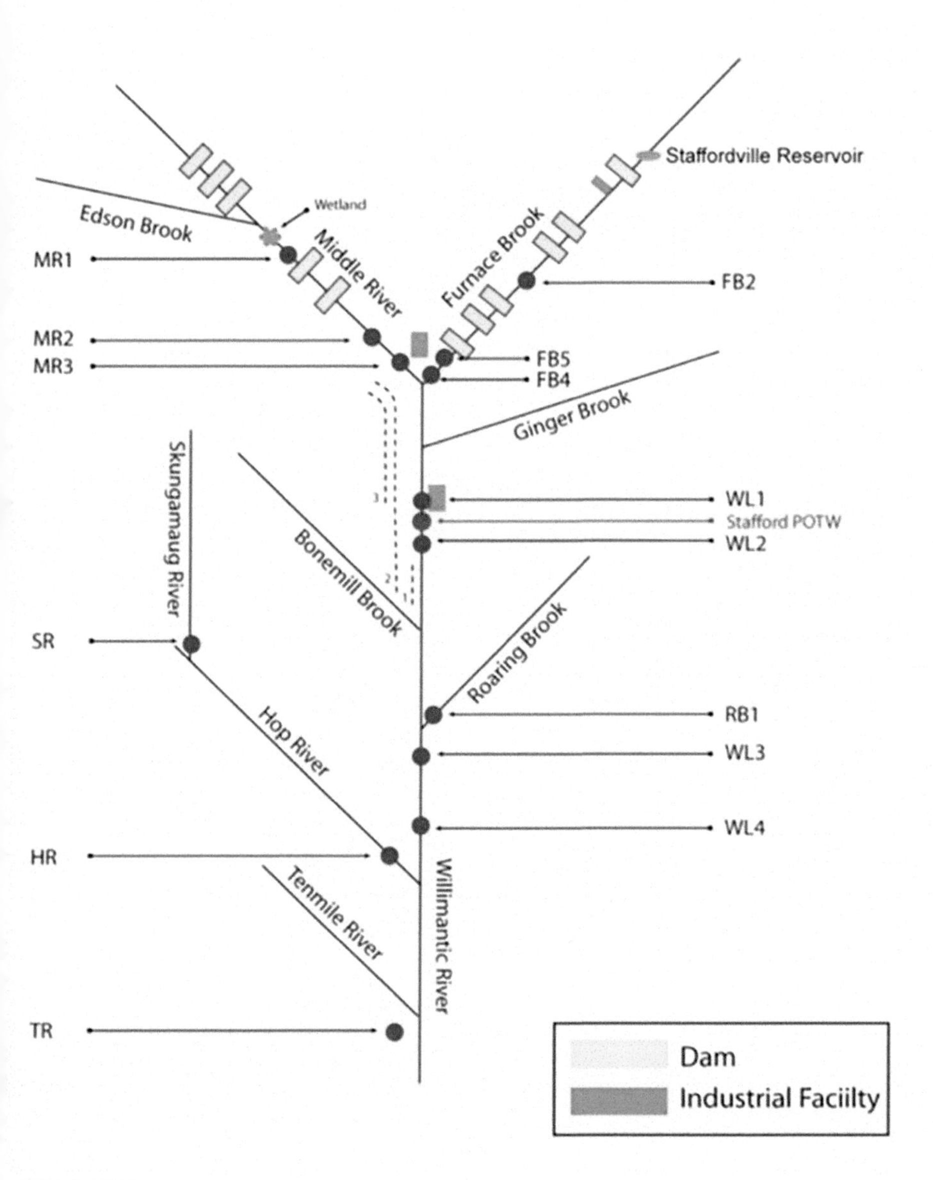

FIGURE 21.1
A diagrammatic map of the Willimantic River, Connecticut. Dots are sampling locations. Broken lines below MR3 indicate sections of river listed as impaired in 1996 (1) and 1999 (2), and the study area (3). MR3, WL1, and WL2 are impaired locations. The unpermitted outfall was between MR2 and MR3.

distribution of effects (Figure 21.1). State biologists waded the reach where the source was thought to occur and found an unpermitted outfall hidden by vegetation. Shutting down that source resulted in recovery of the aquatic community. Presumably, epidemiological assessments of contaminant effects on humans could also benefit from qualitative information from site visits.

21.8 Expert Judgment

When the information that you need is not available and cannot be generated, you can always appeal to expert judgment. In fact, during the course of an assessment, you will make many expert judgments without thinking about them as such. For example, when creating a conceptual model, assessors typically assume that the zooplankters eat phytoplankton. That is clearly true, but it implies an assumption that consumption of particulate organic matter and associated microbes by zooplankters is negligible and can be ignored. Use of a formal process for expert judgment is less common. However, there are many techniques for eliciting expert judgments that attempt to increase the reliability of the results. They include three components. First, define the expertise required and criteria for identifying the experts. Second, define the elicitation process to avoid biasing the results. Third, determine how the results of the elicitation will be analyzed.

Experts are also assembled to develop a consensus concerning major issues in science. The U.S. national organization with that responsibility is the National Research Council of the National Academies of Science, Engineering, and Medicine. They produce more than 200 reports a year including many addressing environmental issues. If you gain some recognition as an environmental assessor, you may be invited to serve on one of their committees. There are other organizations that develop consensus documents at a less lofty level. An example is the Pellston workshops conducted by the Society for Environmental Toxicology and Chemistry.

At best, expert judgment is able to identify scientific consensus on a topic because information exists to be elicited from informed expects. At worst, it elicits guesses from scientists who have little basis for their judgments.

22

Benchmark Derivation

Benchmark values are doses or concentrations of a chemical or other agent that divide acceptable from unacceptable levels. You may use them as:

1. screening values to determine which agents should be considered to have negligible risk and which are contaminants of concern that require further assessment,
2. monitoring thresholds to determine whether environmental conditions constitute impairments,
3. threshold values for regulating the production and use of agents, or
4. permitting standards that separate legally acceptable from unacceptable levels in effluents or ambient exposures.

Benchmarks may be generic or site-specific. A generic benchmark is presumed to be applicable to all or nearly all instances of occurrence of the chemical in a medium or instances of a dose to humans or wildlife. The exceptions should be limited to atypical cases such as saline lakes, streams fed by thermal springs, soils formed from ore-bearing rock formations, or humans with unusual diets or habits. Site-specific benchmark concentrations may be generic values that are adjusted for site chemistry or occurrence of particularly sensitive taxa or life stages (Section 22.5). Site-specific biological adjustments may include removal from the benchmark derivation of data for taxa that are not native to the site or adding results of tests of local taxa. Check the relevant agency's procedures. Fully site-specific benchmarks are derived from tests performed with site species in site media or from observational studies of biotic communities in gradients of contamination levels. For a review of the derivation of ecological regulatory benchmarks, see Belanger et al. (2021).

In addition to doses or concentrations, benchmarks should specify temporal limits. These are typically defined as acute and chronic categories, but durations of exposure and frequencies of recurrence should be specified (Section 4.3).

22.1 A Value and a Factor

The oldest, easiest, and most common method for deriving a benchmark value is to choose a test endpoint for the chemical and divide it by a factor.

DOI: 10.1201/9781003156307-25

The chosen test result may be the lowest value, the one that is most similar to the assessment endpoint, or the lowest value among those that are sufficiently similar to the assessment endpoint. The factor is intended to account for differences between the assessment endpoint and the test endpoint as well as uncertainty concerning those differences (Chapter 26). They are generally called assessment factors.

This method may be applied to whatever data are available or to a prescribed minimum data set. For example, under that European Union's Water Framework Directive, a base data set requires acute test data for a fish, an invertebrate, and a plant. A freshwater ecological quality standard may be derived by dividing the lowest value in an acute base data set by a factor of 1,000 (EC 2011).

Most human health benchmarks such as reference doses are still derived by applying factors to results of a "best" study. The best study is usually the lowest of those that are acceptably relevant and reliable. See, for example, the IRIS human health benchmark values (https://www.epa.gov/iris).

22.2 Methods for Untested Chemicals

Two methods are used to derive benchmark values when a chemical has not been tested for the required endpoint. Read-across extrapolates to the untested chemical from a chemical that is analogous and has been tested for the endpoint. Quantitative Structure–Activity Relationships (QSARs) are statistical relationships between test endpoints or benchmark values and a structural or physical–chemical property. Because the benchmarks are derived without chemical-specific endpoint data, they are usually limited to use as screening values.

Read-across (Section 24.1) is used to estimate a benchmark value for a chemical that has not been tested for the endpoint by extrapolating the endpoint value of a similar but well-tested chemical. For example, dichlorodiphenyldichloroethane (DDD) does not have a human health benchmark value for noncancer toxicity, but dichlorodiphenyldichloroethane (DDT), its parent compound, does have a reference dose, and it is similar to DDD with respect to structure, physical–chemical, toxicokinetic, and toxicodynamic properties (Lizarraga et al. 2019). Hence, by analogy, the reference dose for DDT may be used as an estimate of the reference dose for DDD. This extrapolation method is commonly used for human health assessments but much less for ecological assessments (ECETOC 2012; Rovida et al. 2020; Suter and Lizarraga 2022).

As their name implies, QSARs began as regression models relating toxicity or some other activity of chemicals to their structure, but their independent variables now include physical–chemical properties (Section 24.3). The activity of interest may be a benchmark value or a test endpoint.

22.3 Species Sensitivity Distributions (SSDs)

SSDs are a very common approach for deriving water quality benchmarks. Conventionally, they are distributions of species test endpoints for a chemical which are treated as estimates of the distributions of species sensitivities in biotic communities or assemblages (Chapter 27). That is, they are empirical exposure–response models of communities. For benchmark derivation, they estimate the level of exposure associated with a particular proportion of species affected.

The concentration corresponding to a prescribed centile of a distribution is used as the benchmark value (Figure 27.1). For example, if the 5th centile is used, approximately 95% of species will be protected at the corresponding concentration. SSDs are models of biotic communities represented by a set of species (not randomly selected) that have been tested. SSD-based regulatory benchmarks were first developed by the USEPA (Stephan et al. 1985) and have since been widely adopted (Posthuma, Suter, and Traas 2002). Although the SSD concept is simple, communities treated as a set of individually responding species, there are many issues to be resolved when a nation derives its own version. They include, how many species are required to derive an SSD, what centile will be used to derive the benchmark, will a parametric distribution be fit to the species or will a nonparametric method be used, will representatives of certain taxa be required, will only native species be included, will nonstandard laboratory test results be included, should plants and animals be included in the same SSD, should assessment factors be applied to SSD results, etc.? Greater restrictions on the types of data used and the minimum number of species required result in derivation of fewer benchmarks.

22.4 Weight of Evidence

One common complaint about current practices in benchmark derivation is that they are nearly always limited to standard, laboratory, single-species, toxicity test data (Buchwalter, Clements, and Luoma 2017). To respond to the call for greater diversity, you need a method that can incorporate and integrate diverse evidence. That method is weight of evidence (Chapters 24, 25). The USEPA recommended an unstructured weight-of-evidence approach for developing criteria for suspended and bedded sediment (USEPA 2006). A 2017 draft scoping document for a revision of the U.S. guidelines for ambient water quality criteria for aquatic life included a weight-of-evidence approach. However, the project was withdrawn in response to priorities of a new administration. Two papers that suggest the direction that the USEPA was considering are Suter, Cormier, and Barron (2017) and Suter (2018). An exemplary weight-of-evidence approach was developed by Australia and New Zealand for their *Guidelines for Fresh & Marine Water Quality* (ANZG 2018). A weight-of-evidence approach using field data was derived for fine sediment in New Zealand (Franklin et al. 2019). Two critiques of methods for benchmark derivation disagree on other points, but both concluded that weight of evidence is the best approach (Buchwalter, Clements, and Luoma 2017; Warne et al. 2017).

22.5 Adjustments

When the toxicity of a chemical depends significantly on the chemistry of the medium in which it occurs, benchmark values should be adjusted to apply to the site. U.S. ambient water quality criteria are sometimes adjusted to account for bioavailability (e.g., binding to organic matter) or chemical form (e.g., ionization state). In the U.S., some metals are adjusted for freshwater hardness, some ionizable compounds are adjusted for pH, and ammonia is adjusted for pH and temperature. Some chemicals are adjusted for the presence of sensitive taxa or life stages, such as adding unionid mussels and pulmonate snails to the ammonia criteria (USEPA 2013). Copper is adjusted for multiple water properties to estimate the free ionic concentration, which is the form used in the biotic ligand model of toxicity (USEPA 2007a). Although that mechanistic model is predictive of aquatic toxicity, it has been used by few states because implementing the model at each site is considered to be too complex and difficult. As a result, the USEPA is moving to a multiple linear regression approach that gives similar results (USEPA 2022b). Soil screening values may be adjusted for soil organic carbon–water partition coefficients, air and water diffusivities, water solubilities, Henry's law constants, and pH-specific partition coefficients (USEPA 2005a). Screening benchmarks for organic chemicals in sediment may be adjusted for bioavailability based on partitioning using octanol–water or organic carbon–water partition coefficients (Kow or Koc) (Pluta 2022). Similar adjustments are made by other agencies.

22.6 Levels of Protection

Benchmark values should be set to provide an adequate level of protection. This should include consideration of the duration and recurrence frequencies of exposures, but the major consideration is the magnitude of effect. The threshold magnitude of effect on the endpoint attribute is defined by a magnitude of deviation or the proportion of entities affected. Examples are reduction in growth rate or proportion of organisms dying. In general, 5%, 10%, or even 20% levels are used based on policy judgments supported by scientific judgments.

In the past, particularly for chronic effects, benchmarks were set based on statistical significance of the data for a particular exposure level. It is generally recognized now that statistical significance has nothing to do with biological or policy significance and therefore is inappropriate. However, industry scientists and their consultants still argue for higher levels of permissible effects, such as 25%, based on the magnitude of effects required for statistical significance (Staveley et al. 2018). I have presented the arguments against risk assessment based on statistical significance and responded to Staveley et al. (Suter 2019; Suter 1996) and the American Statistical Association has condemned the use of statistical significance to inform any decision (Section 24.5).

22.7 Presentation of Benchmarks

A hypothetical example of presentation of results for derivation of a water quality benchmark is:

- The quality expressed: threshold dissolved concentration for avoidance.
- The numerical result and units: 1 mg/L.
- Quantitative confidence: 95% CI of ±0.4 mg/L.
- Confidence in the evidence: the body of evidence is moderately relevant and exceptionally reliable.
- Uncertainty: avoidance data are limited to teleost fish. Aquatic invertebrate avoidance (drift) may be more sensitive.

23

Conceptual Models and Network Diagrams

When you assess a complex system, you need to know the components of the system and how they are connected. For contaminants, the basic components are a source, a route of transport, transformation processes, receptor entities that are exposed, and affected endpoint entities. For environmental assessments, these system diagrams are commonly referred to as conceptual models. In other contexts, they have other names such as box and arrow diagrams, causal networks, or system diagrams. They are essential tools of environmental assessment. They are developed in problem formulations of environmental risk assessments (Section 4.2.4).

A critical feature of conceptual models is their representation of causal relationships. Each entity (box) is a cause of the state of entities (other boxes) that are connected (arrow). Each arrow corresponds to a process that somehow translates a change in an upstream entity into a change in the downstream entity. (In graph theory, this relationship is described as parent–child.) For example, Figure 23.1 shows how vacation housing can result in fish kills. Causation is essential to all assessments (Chapter 11) and causal networks help to define the analysis phase (Chapter 5). If your conceptual model is not causal, your assessment can suggest management actions that will not have the intended effect.

23.1 Achieving Shared Understanding

Drawing a network diagram is a good way to reach a common understanding of the system to be assessed. When we began the Bristol Bay assessment, the assessment team met and drew diagrams of the connections between the proposed mining-related activities and components of the watershed's ecosystems. We learned from each other and from the process of organizing available information. We then made each of the component diagrams poster-sized and put them around the walls at a meeting of state, federal, and tribal resource managers. Participants were given markers and invited to add, change, or remove boxes and arrows. As a result, we gained new understanding, improved our conceptual models for the assessment, and gained buy-in of those stakeholders at a fundamental level of the assessment. This use of network diagrams for learning can continue throughout

DOI: 10.1201/9781003156307-26

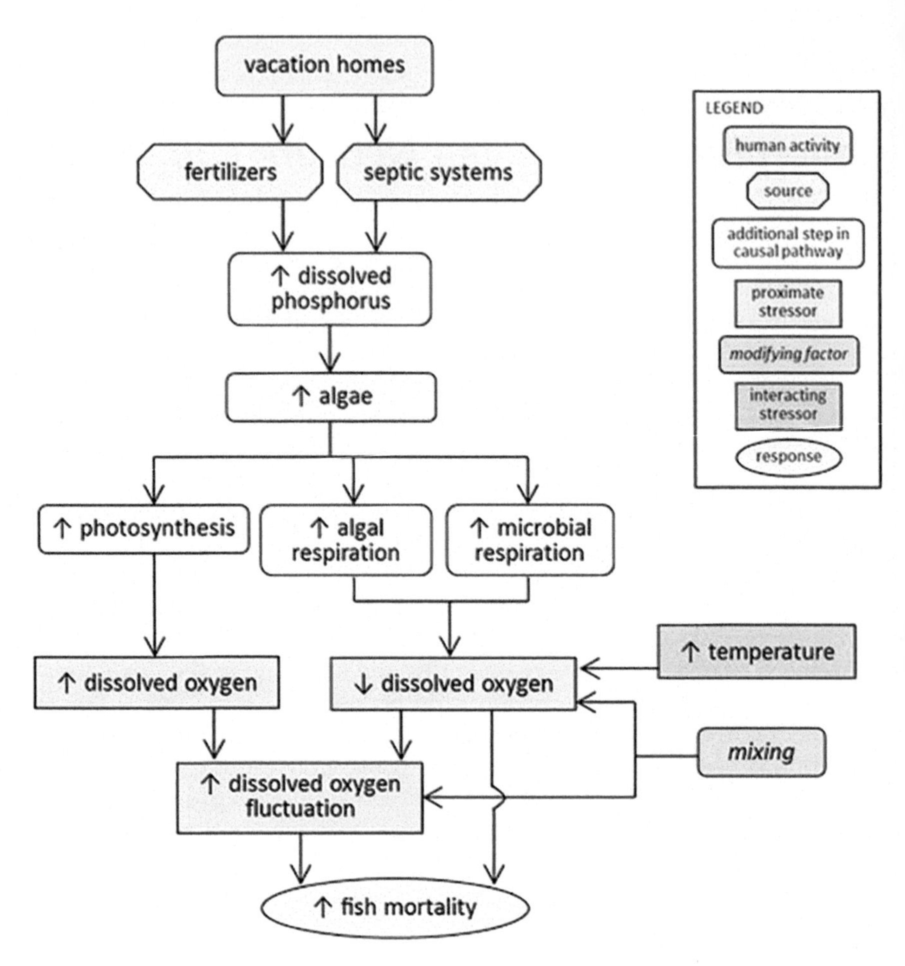

FIGURE 23.1
A site-specific conceptual model for nutrients from vacation homes and fish mortality. A risk assessment would begin with proposed vacation homes and estimate the likelihood or magnitude of mortality. A causal assessment would begin with the fish kill and gather evidence for the causal pathways back to vacation homes. From USEPA (2016e).

the assessment. The analysis and synthesis phases often result in new understanding of the system that prompts modification of the conceptual model.

23.2 Analysis Planning

Planning the analysis phase of an assessment involves determining what must be known and how it will become known (Section 4.2.6). The conceptual model developed during the problem formulation processes provides a foundation for that analysis process. It shows what the assessors understand about the structure of the system being assessed, what information is available about each component of the structure, and what they should observe, measure, test, or model to sufficiently complete their understanding. Hence, the conceptual model will also serve to organize the information as it becomes available.

23.3 Statistical Modeling and Inference

Causal relationships have traditionally been analyzed by unstructured statistical models such as multiple regression. Judea Pearl has become famous for explaining that conventional statistical analyses are not causal, because they analyze associations which are not directional (Pearl and Mackenzie 2018). For example benzene exposure is correlated with and causes leukemia; leukemia is equally correlated with, but does not cause, benzene exposure. Hence, causal models must represent directional relationships in networks which graph theorists call directed acyclic graphs (DAGs).

Consider Figure 23.2. We may know that two environmental variables (A and B) are linearly correlated, but that could be because A causes B (diagram a), B causes A (diagram b), or a third variable C causes both A and B (diagram c). Prior mechanistic knowledge concerning A, B, and C tells us which relationships are true and how a quantitative model should be structured. Further, the structure of the relationships determines causality in ways that conventional statistics does not recognize. For example, if the true relationship is C causes A which causes B (diagram d), multiple regression will not recognize that C has no effect on A's causal relationship with B, because all influence of C is contained in A. Multiple regression will split the regression coefficients between C and A rather than assigning all causality to A. For example, if season, expressed as Julian day (C), is a cause of water temperature (A), which

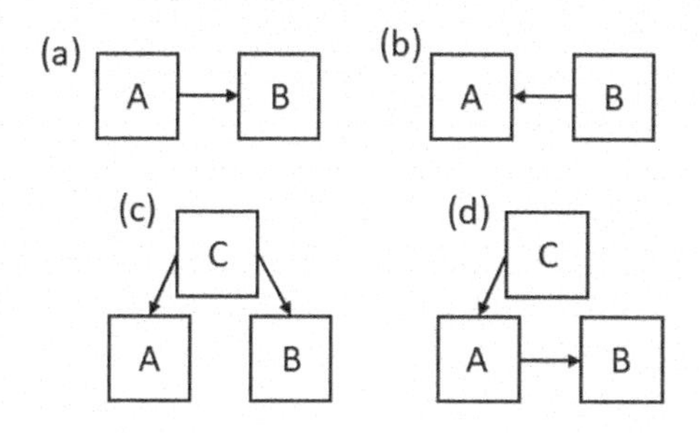

FIGURE 23.2
Diagrams of four potential correlational or causal relationships (a, b, c, and d) among three variables: A, B, and C.

is a cause of growth rate (B), temperature should be included in a regression model of growth rate but not Julian day.

Network diagrams are central to a currently popular method for network modeling, Bayesian networks (BNs) (Section 11.2.2). These models connect the nodes in a network of probabilities of binary or other discrete variables. For example, the probability that fish will die may be 0.2 if below a benchmark for acute lethality and 0.8 if above. Some authors have argued that BNs have improved ecological risk assessments because they are inherently graphical, causal, and probabilistic (Moe, Carriger, and Glendell 2021). They are clearly graphical, but so are other methods such as path analysis, structural equation models, and mechanistic simulation models. I find the other two contentions to be problematic.

Judea Pearl, the developer of BNs, clearly acknowledges that they are only associative and it is the diagram, if anything, that is causal. The structure of the network diagram must be causal in two senses:

1. Represent real-world causation. The network must represent the causal relationships of real-world entities and processes. Pearl provides no guidance on the creation of the diagram itself but states that it is subjective (Pearl and Mackenzie 2018). The diagram comes from the beliefs of the modeler or from the consensus of relevant experts. I suggest that this subjectivism can be minimized by identifying the evidence that supports each causal link in the diagram. If we consider alternative diagrams as alternative causal hypotheses and we define the evidence for each link as was done by Norton and Schofield (2018), then we can weigh the evidence for each hypothetical diagram.
2. Represent the unconfounded relationship of the cause of interest to the effect. Even after we have a network diagram that is directional and well supported by evidence, we must assure that the analysis will correctly determine the strength of the causal relationship of interest. That is, even accurate diagrams can be confounded, and "surgery on models" must be performed to make the modeled network resemble an experiment or a counterfactual statement (Chapter 11). Only then will the model derive an unconfounded estimate (Pearl and Mackenzie 2018). Unfortunately, some environmental scientists state that BNs are inherently causal. As a result, they create BNs without doing the work of making them truly causal.

BN models are said to improve environmental risk assessments by being probabilistic (Moe, Carriger, and Glendell 2021). The authors justified the need for probability by citing USEPA and EFSA definitions that actually do not refer to probability but only to "likely" and "likelihood." It is clear to me that these definitions are referring to the ordinary English usage of those terms. The agencies certainly did not mean to define risk as the probability of data given a hypothesis (the sense in which likelihood means a type of probability). Otherwise, why did the USEPA and EFSA not say probability if that is what they meant? In any case, endpoints expressed as probabilities are potentially confusing, not necessary, and generally not desired by decision-makers (Chapter 18).

There is also a tendency to prefer BNs because Bayesian probabilities are subjective (Box 18.1). I have been an observer of one BN modeling workshop and a participant in another. In both cases, the experts had little hesitancy in guessing probabilities without data or evidence and without concern for confounding or other issues with the structure of the diagram. The probabilities in BNs run into the problem of vagueness discussed in Chapter 18. BNs for environmental problems have described the probabilities variously as uncertainties, frequencies, weights of evidence, judgments, or just as probabilities. In some applications, it is clear from the justifications that the probabilities for different nodes have different interpretations, such as frequencies when data are available and uncertainties when data are not.

An alternative approach to statistical modeling of causal networks is path analysis, developed by the great population geneticist Sewall Wright, and its descendant, structural equation modeling (SEM). SEMs, in contrast to BNs, are not subjective. They are data demanding, although they can handle some latent (unmeasured) variables (Shipley 2000).

So far, both BNs and SEMs have been used in the USEPA primarily in a research context (Malaeb, J K. Summers, and B. H. Pugesek 1999; Carriger and Barron 2020). However, in at least one case, network modeling has proven to be useful to a program. The USEPA has struggled for decades to develop water quality criteria for nitrogen and phosphorus. The problem is that these are important nutrients, but above some level, they cause excessive algal production, including cyanobacteria and their toxins which cause direct effects, which in turn cause endpoint effects. The newest criteria are based on BNs of the network of relationships between N and P input and the drinking water, recreation, and aquatic life endpoints (Figure 13.1). The Bayesian analysis is based on relationships between variables such as organic matter concentrations and deepwater dissolved oxygen. The probabilities are categorical relative frequencies and the probability distributions are used to create credibility bounds (the Bayesian analog of confidence bounds) that define the protective criteria (USEPA 2021a). Hence, the BN is a way to derive a protective benchmark based on a high probability of protection equivalent to the 99.9th centile in a Monte Carlo analysis of dietary exposure (Section 18.2). Like other benchmark derivations, it works back from a threshold level of effect to the benchmark level of exposures, the nutrient concentrations (Chapter 22).

23.4 Mathematical Modeling

Conceptual models began as tools for mathematical simulation modeling. Before writing equations, modelers draw diagrams in which the boxes represent state variables, such as the contaminant concentration in the first fully mixed stream reach, and the arrows into that box represented processes, such as stream flow and effluent flow. Once the conceptual model is completed, the implemented model is created by writing the equations for the processes and by determining values of parameters. Simulation modelers still use conceptual models this way, but their equations may not be limited to old-fashioned linear differential equations. A good example is a model of mercury sources, transport, fate, and effects in the Everglades ecosystem (Pollman, Rumbold, and Axelrad 2020).

23.5 Adverse Outcome Pathways

Adverse outcome pathways (AOPs) are conceptual models developed to represent toxicology at the molecular level and potentially extending to population attributes (Ankley et al. 2010). They begin with a molecular initiating event such as membrane ligand binding, DNA binding, or hydrophobic interaction with cell membranes. They end with an adverse outcome, which may be a conventional assessment endpoint but more often is a suborganismal toxic effect (Figure 23.3). The AOP links the molecular initiating event and adverse outcome through series of causally connected key events. The causal links are termed key event relationships. A peculiarity of AOPs relative to other conceptual models is that they are agent-agnostic. Any chemical that

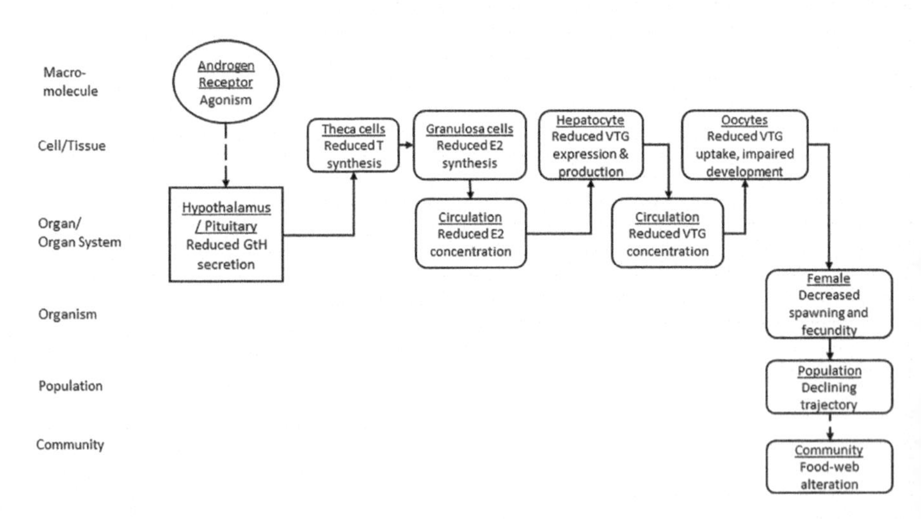

FIGURE 23.3
An adverse outcomes pathway (AOP) for androgen receptor agonism leading to reproductive dysfunction in repeat-spawning fish, by Dan Villeneuve, USEPA (https://aopwiki.org/aops/23). 17ß-trenbolone is a prototypical agent for this AOP. Unlike most AOPs, this one is extended to effects at high levels of organization.

causes the molecular initiating event will set in motion the sequence of key events, like falling dominos, leading to the adverse outcome. Hence, AOPs are conceptual models of mechanisms that can apply to numerous chemicals. The point of this relatively new concept and associated terminology is to logically assemble, interpret, and communicate the flood of data generated by in vitro tests and relate them to the in vivo test results that represent endpoint effects. Unlike other conceptual models, AOPs have standards and platforms for developing and sharing (OECD 2022a).

23.6 DPSIR

DPSIR, the Drivers, Pressures, State, Impacts, and Response model, is a conceptual model developed for impact assessments that represents societal and environmental linkages (Figure 23.4). It is popular in the European Union where it provides the conceptual organization for the Water Framework Directive and other regulations (EC 2003a). The steps with respect to water pollution are:

Drivers are the sources of the problem such as tilled agriculture.

Pressures are the pollutants added to the water such as silt, fertilizer, and pesticides.

State is the physical and chemical condition of the water.

Impacts are effects on aquatic life and human health.

Response is the regulatory and management reactions to the State and Impacts aimed at modifying the Drivers, Pressures, and States to reduce the Impacts.

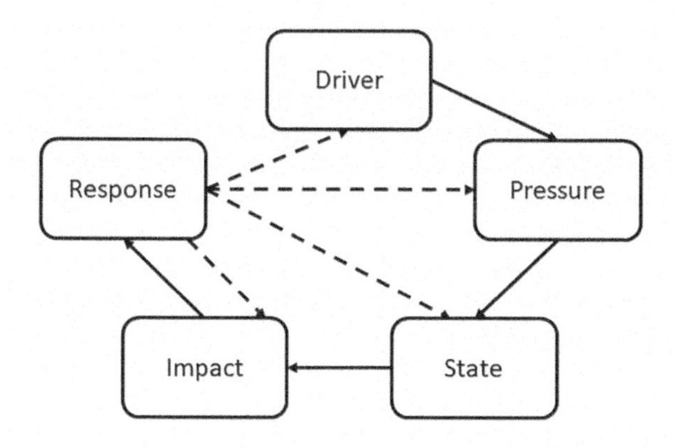

FIGURE 23.4
The driver, pressure, state, impact, response (DPSIR) framework for environmental assessment in a social-political context. Solid lines indicate the causal processes and dashed lines indicate feedback from the response.

DPSIR can serve to clarify the scope of an environmental assessment. For example, many environmental assessments are limited to relating the state of a system to the impacts. After impacts are shown to be significant, the drivers (sources) and pressures (releases) may be considered in a causal assessment. Responses may be determined by a separate team that includes engineers and attorneys.

23.7 Strategies for Conceptual Models and Other Causal Diagrams

Because of the importance of conceptual models to environmental assessments, you must assure that they are scientifically correct, and you should design them to be clear and readily understood. The scientific defensibility of a conceptual model results from its consistency with evidence and peer review. Some processes and relationships are self-evident and need no justification; water flows down-gradient and soluble chemicals are diluted by mixing. However, others such as biomagnification of an organic chemical or population declines caused by endocrine disruption should be supported by evidence. Some components of your conceptual model may be only hypothetical during the problem formulation and must be justified or altered after the analyses and synthesis are performed. Peer review of a conceptual model need not be a separate formal process. The involvement of an assessment team and stakeholders in the development of a conceptual model may be sufficient to judge its scientific defensibility.

Some network diagrams resemble the web of a drunken spider. To avoid confusion within your assessment team and enhance communication with others, you should logically design the network. A common structure for pollutants is a flow of processes from release at the source, through transport and fate in the environment, to uptake and induction of effects in human and nonhuman receptors. These may be arranged from top to bottom or left to right. You should label the boxes with the represented entities, and, at least for mechanistic models, label the arrows with the process by which one entity affects the other. For example, Figure 23.5 shows a conceptual model that can not only show the causal relationships among entities (down arrows), but can also identify the functional relationships that are used to calculate the strength of those causal relationships.

Complex conceptual models may have so many boxes and arrows that the logic gets lost. Such complexity occurs more often in impact assessments for projects than in risk assessments for pollutants. For example, infantry training involves many activities, one of which is tank maneuvers. Those maneuvers in turn include disturbance and compression of soil, crushing of organisms, behavioral disturbance of wildlife, and refueling spills. Each of

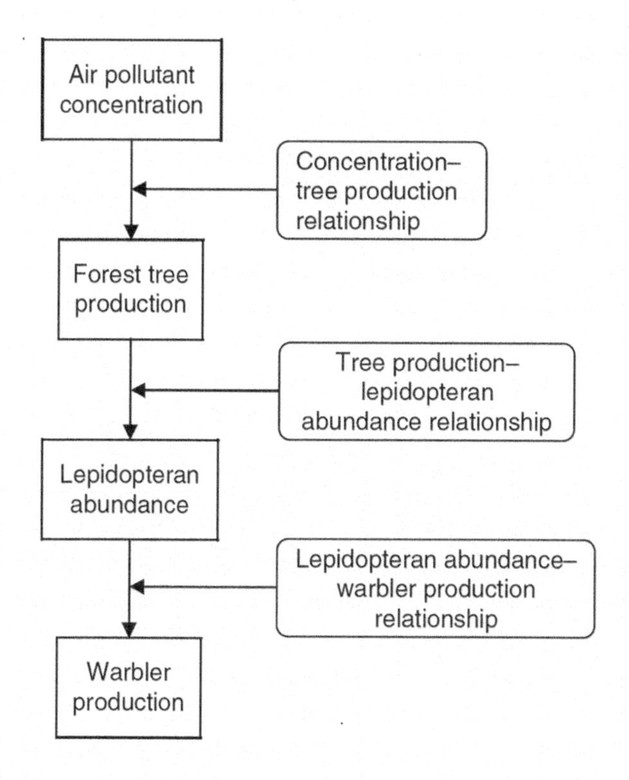

FIGURE 23.5
A simple conceptual model of the effects of air pollution on warbler production, showing one way in which the variable entities and the exposure–response relationships that would be used to determine the results of the connections can be displayed.

those may be diagrammed as a module, and the connections among the modules leading to endpoint effects may be diagrammed to clarify the combined causal relationships (Suter 1999). In other words, they are best diagrammed as nested hierarchies.

23.8 Alternative Conceptual Models: Maps and Pictures

Diagrams of causal networks are the most generally useful conceptual models, but pictorial and cartographic conceptual models are useful in site-specific assessments. Maps may represent the locations of sources and receptors and the pathways in between. They may also clarify the locations along pathways of water, sediment, soil, and biological samples that have been chemically analyzed, and of biotic communities that have been identified and quantified. The diagrams and maps may complement each other during problem formulation by ensuring that all causal pathways are considered.

Spatial conceptual models can also aid inference. The first application of the stressor identification guidance (USEPA 2000b) was on the Willimantic River in Connecticut (Bellucci, Hoffman, and Cormier 2010). Mapping the data on a diagram of the river showed that impairment of the benthic invertebrate community was greater upstream than downstream of the wastewater treatment plant, which was a suspected cause of impairment. The overall pattern of impairment suggested a source on an upstream tributary (Figure 21.1). A field investigation found a hidden, unpermitted, episodic, and toxic discharge just above the worst-impaired site (MR3).

Maps showing causal relationships may aid communication. When we were assessing the risks from contaminants from the Oak Ridge Reservation, a doctor insisted that the undiagnosed complaints of his patients were due to contaminants from the Oak Ridge Reservation in the city drinking water supply. A map would have been useful to show that city drinking water came from the Clinch River and that all flows off the reservation and into the Clinch River were well downstream of the drinking water intake. Simply asserting that fact at a public meeting was not convincing to some members of the public, and a map showing flows of drinking and contaminated waters would likely have helped.

Pictorial conceptual models are also primarily a communication tool. They pictorially portray a scene with media, biota, and humans connected by arrows representing contaminant transport, uptake, and direct and indirect effects. They are believed to be more appealing to the public and other non-scientists than a box and arrow diagram, but they are less clear, and the ambiguities greatly reduce their utility for assessment planning, inference, or modeling. Does a pictured bird represent a specific taxon of birds, a trophic group, all birds, or all vertebrates? What mode of exposure does the arrow to humans represent?

23.9 Conclusions

Diagrams of the causal relationships in the system being assessed are remarkably versatile and useful. One of my junior colleagues told me that she was not a graphical person. However, after a year or two, she informed me that she was a convert and used conceptual models frequently to work out her approach to assessment problems. So, if you have an aversion to network diagrams, give them a chance.

24

Inference

True genius resides in the capacity for evaluation of uncertain, hazardous, and conflicting information.

—*Winston Churchill (1874–1965)*

Inference is reasoning that converts data and other information into evidence that can support or discredit a qualitative or quantitative hypothesis. For example, the body of evidence supports the hypothesis that low dissolved oxygen caused the fish kill and that the kill occurred where levels were below 2 mg/L. There are three types of inference.

Deduction is inference from a general rule to a specific conclusion. For example, water flows downhill in response to gravity (a general rule). Therefore, contaminated water upstream of a release of aqueous contaminants is not due to that source (a conclusion). Deduction is reliable if the rule is reliable.

Induction is inference from evidence derived from specific observations to a general conclusion. For example, from the result of a toxicity test (a specific observation), you may inductively derive a general conclusion that the chemical is not lethal to the tested species below an observed no-effect concentration of 1 mg/L. Induction is not logically reliable, because other cases may be different from the observed case. For example, the toxicity in the stream that you are assessing may be different because the dilution water chemistry in the field is different from laboratory water.

Abduction is inference from incomplete information to the best explanation of the evidence. It is a type of inference commonly employed by scientists, diagnosticians, detectives, and assessors. For example, a chemical has been released at an outfall at lethal concentrations in the past. The best explanation of a new fish kill below the same outfall, given the history, is that it was caused by that same effluent and chemical. Abduction is commonly used for hypothesis formation, as in this example. However, weight of evidence can be used to abductively reach a conclusion with sufficient confidence to determine a decision.

In this chapter, I briefly describe the major categories of inferential methods commonly used in environmental assessment. In the end, most of your inferences by any of these methods will be some variant of abduction.

DOI: 10.1201/9781003156307-27

24.1 Analogy

Analogy is an inferential method based on the assumption that if *A* is similar to *B*, then *A* is likely to have properties of *B*. Analogy is sometimes criticized as a weak basis for inference. For example, the eminent epidemiologist K. Rothman condemned A.B. Hill's inclusion of analogy among his causal considerations. Rothman argued that analogy cannot be trusted, because it is limited only by a person's imagination (Rothman and Greenland 1998). However, analogy is a useful tool if either (1) its use is limited to inspiration or (2) it is used to draw conclusions guided by a clear structured method rather than free imagination.

Spontaneous inference by analogy simply uses similarities to stimulate inspiration and suggest hypotheses or ideas for research or analysis. For example, you might say, "this stream reminds me of a stream that I assessed in the past, and that one was impaired by acid mine drainage, so this one is likely to be impaired by acid mine drainage too." That analogy would lead you to gather data that could provide evidence for that cause. This use of analogy has no method; it often results from unconscious mental processes. That is, you look at a stream and a previously studied stream springs to mind.

Formal inference by analogy in environmental assessment is best represented by read-across (Section 22.2). For many chemicals, there is little toxicology or environmental chemistry information. Read-across solves the missing information problem by a process of finding a well-characterized chemical (the source chemical) that is an analog of the incompletely characterized chemical being assessed (the target chemical) (ECETOC 2012; Rovida et al. 2020; Suter and Lizarraga 2022; Lizarraga et al. 2023). Structural, physicochemical, toxicokinetic, and toxicodynamic properties that are known for both the target and a potential source chemical are compiled and compared. A common property of the chemicals such as molecular weight or octanol–water partitioning coefficient is evidence for or against toxicological similarity, depending on how similar the properties are in the two chemicals. Then the overall weight of evidence for similarity of the chemicals is determined based on the number and diversity of properties that are similar or dissimilar. The potential source chemical that is most similar to the target chemical is identified as the analog. If that most similar chemical is sufficiently similar to the target chemical, it is considered to be a source chemical, and its quantitative and qualitative properties can be applied to the target chemical.

Another example of inference by analogy is estimation of background values for one poorly characterized area from the well-characterized background value for a similar area (Cormier et al. 2018). The background specific conductivity for streams in the portion of ecoregion 70 in Ohio cannot be well characterized from existing specific conductivity data, but the portion in West Virginia is well characterized. We weighed evidence of similarity of properties of the ecoregion in the two states that are relevant to specific conductivity. Twenty-seven types of evidence belonging to four categories of evidence (physical properties, measured specific conductivity, spatial distribution of low SC sites, and biological properties) were identified and weighted. Because those types of evidence were similar in the two states, we inferred by analogy that West Virginia background values could be used for Ohio.

Objections to the use of analogy still occur, but they are mainly applied to uses like Hill's which do not include a structured method for making the inferences. Structured methods make inference by analogy transparent and defensible, but of course they are still not as reliable as inference from the desired data.

24.2 Screening

Screening is a process of elimination intended to identify agents, media, or sources that are either clearly harmless or clearly unacceptably hazardous. Screening is a simple process that narrows the assessment to considering only a few ambiguously hazardous agents. It may be part of a tiered assessment scheme (Chapter 4) or a single preliminary step in an assessment. A screening assessment compares an exposure estimate to a screening benchmark value (Chapter 22). In cases of direct exposure to the contaminated medium, such as a fish in water or a worm in soil, the conservatism of exposure comes from assumptions such as exposure to the maximum measured concentration. If exposures are doses, simple models such as food ingestion rates provide doses to be compared to benchmark doses. An example for human health assessments in the U.S. is the regional screening values (USEPA 2022d). Screening values for some chemicals are available for indoor and outdoor air, tap and ambient water, sediment, fish flesh, and soil. In cases that involve complex sources and routes of transport and exposure, a generic screening model of sources, transport, and fate may be used to estimate exposures that are compared with screening benchmarks. An example is the USEPA's Risk-Screening Environmental Indicators (RSEI) model for human health risks for industrial complexes (https://www.epa.gov/rsei). It is used for prioritization of sources or contaminants based on the toxics release inventory. Ecological soil and sediment screening values are also provided by the Agency (USEPA 2005a; Pluta 2022). The soil, water, sediment, and wildlife dietary screening values created by my group at ORNL are still used, but they are out of date and should not be used without updating (Sample et al. 1998).

24.3 Empirical Estimation

Empirical estimation is the derivation of a value from data (as opposed to judgment, mechanisms, or theory). The estimate may be a summary statistic such as a mean or median, preferably with an estimate of confidence. It may also be a regression model or other statistical model of relationships in the data. It can be used to estimate parameters, exposure levels, effects, or exposure–response relationships.

The calculation of a mean and confidence interval from results of a study should be familiar to all of you. The quantities might be endpoint exposures or effects levels or a characteristic such as water solubility. This may seem simple and traditional, but emphasizing estimation and confidence rather hypothesis testing has been termed "the new statistics" (Cumming 2012; Cumming and Calin-Jageman 2016).

It is desirable, but uncommon, to present multiple confidence intervals and not just the conventional 95% interval which has no significance for environmental management other than familiarity. The use of 95% confidence is a carryover from Fisher's significance level, but confidence and significance are not the same and estimation and hypothesis testing have different purposes. USEPA guidance requires that risk characterizations include confidence in the results (Chapter 8).

When multiple studies of the same type are available, you may use a technique called meta-analysis to combine results into a single estimate and confidence interval (Cumming 2012). If all of the studies have acceptable quality, the mean of their weighted results should provide a better estimate than any one study and their confidence interval should be narrower than any one study. Exceptions may occur if, as with any empirical method, some studies have unrecognized flaws. Meta-analysis is more common in human health because multiple studies of the same human hazard or treatment are common. However, some ecological issues such as trends in insect abundance allow meta-analyses because studies of the question are numerous (Section 21.2). Meta-analyses are typically conveyed by forest plots that plot the mean and confidence interval for each study and for the combined estimate. In sum, meta-analysis should provide the best estimate of the endpoint, and its confidence interval should convey the confidence provided by replication.

You may also derive estimates from empirical models of relationships. The most prominent empirical models in environmental assessment are exposure–response relationships (Figure 7.1). Conventionally, statistical functions are fit to data from toxicity tests, and the parameters are derived by least squares regression. However, maximum likelihood, nonparametric, meta-regression, or other methods may be used. The derived model may be used to estimate the effects of a particular exposure level (i.e., the solution Y of the model solved for an X value). It may also be used to estimate the exposure level corresponding to a particular effect (i.e., the solution X of the model solved for a Y value). You need to estimate the effects (Y) for a risk assessment and the exposure (X) for deriving a benchmark value or in causal assessments to determine the exposure sufficient to cause an observed effect. Whether predicting effects or exposure levels, you must justify the use of a particular exposure–response relationship to infer your endpoint. That is, how relevant is the study or why is the relationship believed to be predictive and causal? Guidance for exposure–response modeling is abundant, including the USEPA's benchmark dose software (https://www.epa.gov/bmds). Returning to the new statistics, meta-regression provides an integrated regression model by combining multiple studies.

Another important type of empirical model is Quantitative Structure–Activity Relationships (QSARs). These are models that relate a chemical's activity of interest to a structural or physicochemical property. Activities of interest may include a toxic response, an uptake rate, or a partitioning coefficient between two environmental media. The predictor property should be predictive of the activity of interest and should be known for most or all chemicals (e.g., molecular weight) or should be much quicker and easier to generate than the property of interest (e.g., bioconcentration factor). QSARs are particularly important in the assessment of individual chemicals under TOSCA and REACH (Chinen and Malloy 2022).

Classic ecotoxicological examples of QSARs include the use of octanol–water partitioning coefficients to estimate bioconcentration factors and acute lethality of organic chemicals (Konemann 1980; Veith and Kosian 1983; Veith, Call, and Brook 1983). QSARs may be purely empirical. However, others such as Veith's model for estimating acute lethality to fathead minnows from octanol–water partitioning coefficients are based on assumed mechanisms. Organic chemicals that fit this model are acting by baseline narcosis. Chemicals that are more toxic than baseline narcosis have a more specific mechanism of action.

To develop a QSAR model, you must identify a category of chemicals that are thought to have common activities as a result of their structures or physicochemical properties. Then try out different relationships between your activity of interest and potential predictor properties. There are several computer apps that will do that for you and evaluate the potential QSARs by cross-validation or some other criterion. Once you have a good model, you can use it to quantitatively estimate the property for chemicals that have not been tested or measured. However, in many cases, you can use an existing QSAR. The best current compilation of QSARs and guidance on their use is the OECD and ECHA toolbox (OECD 2021).

24.4 Estimation by Mathematical Simulations

Mechanistic modeling derives estimates by mathematically simulating the processes that generate the desired endpoint value. The most common and generally accepted mechanistic models in environmental risk assessment are those that simulate the transport and fate of chemicals. For example, the USEPA (2022f) provides standard models and guidance for atmospheric modeling including dispersion, photochemistry, and receptor modeling. Mathematical models of effects are uncommon in ecological and human health assessments. The human health exception is toxicokinetic models, which are models of the fate of chemicals in the body. The ecological exception is demographic models that are used to estimate effects on important nonhuman populations. These models convert estimates of effects on survival, development, and reproduction into estimates of effects on population size and rates of population growth or decline.

Although these models are mechanistic, they often have empirical elements because a parameter or functional form is unknown. The process of fitting to data concerning past conditions or conditions in a similar system is termed model fitting or tuning. Hence, although the equations of a mechanistic model may be scientific rules from which the result is deduced, a complete environmental model is typically a mixture of mechanistic and empirical components. Generic guidance on developing and applying environmental models is provided by the USEPA's Council on Regulatory Environmental Models (USEPA 2009).

The mechanistic mathematical models discussed above are systems of multiple equations, but simple mathematical models are useful and easily implemented. For example, a simple equation estimates the concentration of a chemical in an effluent fully diluted in a stream. Another simple equation estimates the dietary exposure of an animal consuming a contaminated food item. I will not provide them here because you should be able to figure them out yourself. However, I have been asked for both of those equations by graduate students. If you look at the units of dietary dose (mg of chemical per kilogram of animal per day), the equation practically writes itself.

24.5 Hypothesis Testing

Statistical hypothesis tests are intended to determine whether to accept or reject the truth of a hypothesis based on data analysis. For example, in a toxicity test, you may hypothesize that an exposure to a chemical has an effect on a test species relative to no exposure. You may hypothesize that in a toxicity test, more mice die following exposure (the alternative hypothesis H_1) than die when unexposed (the null hypothesis H_0). Statistics have been used to determine whether differences in mouse mortality are "significantly" more likely to be due to exposure than to chance. Similarly, field studies may test the hypothesis that contaminated biotic communities have fewer species than uncontaminated communities. Although the concept of statistical significance has been widely rejected, it is still widely used.

Nearly all of us have been trained in frequentist statistical hypothesis testing. In its standard form, we assume that the null hypothesis (H_0) is true and calculate the likelihood that data as extreme as ours or more extreme will occur by chance given that there is no effect. Conventionally, the standard for rejecting H_0 is a p value of 0.05 or less. In other words, there is no more than a 1 in 20 chance that our effects data differ from zero effect due to random variance. Many statisticians and scientists have criticized this approach to inference. I wrote an extended explanation of why it is wrong for environmental assessment, and recently, the American Statistical Association determined that it should not be used as a basis for decision making (Suter 1996; Wasserstein and Lazar 2016). However, this statistical significance approach is still routinely used to determine whether a scientific finding will be accepted. Wasserstein and Lazar's (2016) pathetically humorous circular explanation is:

> Q: Why do so many colleges and grad schools teach $p=0.05$?
> A: Because that's still what the scientific community and journal editors use.
> Q: Why do so many people still use $p=0.05$?
> A: Because that's what they were taught in college or grad school.

An alternative is Neyman-Pearson statistics which are based on error rates which must be balanced. You probably know it from discussions of Type I and Type II errors with corresponding error rates signified by α and β. It requires that the acceptable error rates and effect size be specified in advance, but that creates problems in practice. For example, the Data Quality Objectives process for designing sampling programs at Superfund sites required specifying the acceptable limits of decision error (i.e., α and β) (Quality Assurance Management Staff 1994). I happened to be the ecological lead on the first assessment (the Clinch River remedial investigation) to try out the process for a real site assessment. It seemed reasonable that a decision-maker would be the one to balance false positive and false negative error rates for declaring a site clean enough against the costs of sampling and analysis to reduce error rates. The decision-maker from USEPA Region 4 would not specify that any decision error rate was acceptable. Despite that problem, error statistics are still used to design contaminated site sampling and characterization of contamination in the ProUCL statistical software (https://www.epa.gov/land-research/proucl-software). It is treated as a sampling optimization process rather than as a policy decision.

An increasingly popular approach is Bayesian hypothesis testing which uses a ratio of likelihoods for two hypotheses called a Bayes factor. Commonly, the likelihood of a hypothetical effect is compared to the likelihood of a hypothesis of no effect so that the Bayesian test is equivalent to the conventional frequentist null hypothesis test. It has the advantage that the null hypothesis is not assumed to be true. If the Bayes factor is less than 1, that indicates that the test supports the null (no effect) hypothesis. If it is more than 1, that indicates that the test supports the alternative (real effect) hypothesis. It has the additional advantage that Bayes factor is not used to

TABLE 24.1

Interpretations of Bayes Factors for an Alternative Hypothesis (Kass and Raftery 1995)

Bayes Factor	Evidence against the Null Hypothesis
1–3.2	Not worth more than a bare mention
3.2–10	Substantial
10–100	Strong
>100	Decisive

distinguish true from false hypotheses. Rather, the factors are treated as defining a gradient of the strength of evidence for the hypothesis of interest relative to some alternative hypothesis (Table 24.1). An advantage of Bayes factor relative to other Bayesian statistics is that it does not require prior beliefs (Box 18.1).

Relative risk, likelihood ratios, and odds ratios may be used similarly to Bayes factors in hypothesis testing. They are used as summary statistics for frequencies as in Chapter 8. For testing, they must be assigned interpretations like those in Table 24.1. The interpretations should be based on the quality of the study as well as the result. An effect in a randomized controlled toxicity test might be considered real even at a small relative risk such as 2. A relative risk of 3 might be sufficient to conclude that an effect is real in a good quality observational study. The greater the ratio, the greater your confidence, but you should also consider the frequencies. If the mortality is 1 of 100 in the controls and 2 of 100 in the treatment, that is a doubling, but it is just a 1% increase in dead organisms. To avoid biased interpretations, you should establish your criteria in advance. Such analytical results and interpretations are not commonly used to express results in environmental studies, but could be better for hypothesis testing than p values.

Hypothesis testing has become almost an inevitable form of inference in the natural sciences, and it may be expected by your decision-maker or a stakeholder. However, estimation of effects levels and associated confidence is more useful in most situations. When evidence for a hypothesis is multiple and heterogeneous, weighing of evidence is better than hypothesis testing. As the American Statistical Association concluded, overall scientific reasoning should prevail over a hypothesis test result (Wasserstein and Lazar 2016).

24.6 Weight of Evidence

The concept of weight of evidence comes from the ancient Greek goddess of justice, Themis, who weighs the evidence on each side of a legal case. She decides, based on the relative evidential weight in the pans of her scales of justice, for or against guilt of an accused person or for one person or the other in a civil dispute. The distinguishing feature of this inferential method is that the evidence may be quite diverse. In a criminal case, the evidence may include testimony of the accused, victim, or witnesses; documentary evidence of motive; testimony of forensic scientists; and others. Although individual pieces of evidence can be quantified, clearly such bodies of diverse evidence cannot be quantified and statistically analyzed to determine the judgment. Many of our inferences in environmental assessment are similar to Themis's legal inferences. A legal definition of WoE is "The degree to which evidence convinces triers of fact to either accept or reject a factual assertion" (Legal Information Institute 2022). A biological definition is analogous: "Weight of evidence refers to a systematic approach that scientists use to evaluate the totality of scientific evidence to assess if the science supports a particular conclusion" (Endocrine Science.org 2022). WoE has become essential to environmental assessment. As the Canadians say with respect to their Environmental Protection Act, "The nature and scope of risk assessments may vary, but all have a WoE component" (Health Canada 2022).

The traditional method for weight of evidence is narrative. Like an attorney presenting a case or a scientist publishing a review, an assessor inferring an attribute or effect of an agent will present the evidence and a narrative argument for a hypothesis such as the eutrophication is caused by phosphorus from fertilizer. The inferential task of narrative weight of evidence is to make sense of the evidence and present the results of sense-making in a clear and convincing account. The flaw in this method is the narrative fallacy. That is, the true inference is not necessarily the best story, and a talent for spinning random evidence into a plausible narrative does not make you a good assessor.

Narrative weight of evidence may be most useful when the inference is simple and can be explained in a sentence of two. However, a good narrative may be needed to convey a complex causal inference with abundant evidence. A 25-year study of the cause of vascular myelinopathy in bald eagles and other species was resolved by combining several individual investigations into a narrative. The story involves an introduced aquatic plant that provided a substrate for a cyanobacterium that produced a toxin when sufficient boron was present which was supplied by geologic or anthropogenic sources (depending on the watershed), and ultimately, the eagles were exposed by consuming prey that were debilitated after consuming the plant covered with cyanobacteria (Breinlinger et al. 2021; Stokstad 2021). The narrative is convincing in this case because numerous case-specific empirical and mechanistic studies were done that characterized the steps in the causal pathway.

In contrast, multiple reviews of empirical studies have addressed the causes of decline in birds by using the narrative that pesticides are causing insect abundances to decline which causes bird abundance to decline. Hallmann et al. (2014) showed co-occurrence, addressed two confounders, and concluded that bird declines were associated with imidacloprid use via insect declines. Bowler et al. (2019), in contrast, associated bird declines with agricultural intensification and particularly loss of grassland habitat. Tallamy and Shriver (2021) associated bird declines with insect declines but not any particular cause of insect declines. None of the three papers set criteria for demonstrating causation or explicitly weighed the evidence, and the narratives were appealing but not convincing.

A better approach is the use of evidential criteria. This involves defining a set of types of evidence that can provide a clear result for a particular type of inference. An example is Koch's postulates for determining that a pathogen causes a disease. These postulates have been adapted to chemical effects on humans (Yerushalmy and Palmer 1959; Hackney and Linn 1979), air pollution effects on crops and forests (Woodman and Cowling 1987; Adams 1963), and ecological causal assessment in general (Suter 1990, 1993b). My version of Koch's postulates for environmental assessment is:

1. The injury, dysfunction, or other putative effects of the agent must be regularly associated with exposure to the agent and any contributory causal factors.
2. Indicators of exposure to the agent must be found in the affected organisms.
3. The effects must be seen when organisms are exposed to the agent under controlled conditions, and any contributory factors must contribute in the same way during the controlled exposures.
4. The same indicators of exposure and effects must be identified in the controlled exposures as in the field.

This rather strict set of causal criteria is explained and an example use is provided on pages 55 and 56 of Suter (2007). An independent but conceptually equivalent approach is the sediment quality triad (Chapman 1990). It requires that: (1) the sediment must be analyzed and the results compared to benchmark concentrations for the detected chemicals, (2) the sediment must be tested for toxicity, and (3) the contaminated sediment biota must be compared to the biota of reference sediments. Sets of evidential criteria must, in most cases, be satisfied by case-specific studies.

A much more flexible approach to weight of evidence is the considerations developed by the U.S. Surgeon General's Commission and A.B. Hill to determine whether the association of lung cancer with cigarette smoking is causal (Hill 1965; USDHEW 1964; Suter, Nichols, et al. 2020a). The considerations are a mixture of types of evidence, characteristics of causation, and properties of pieces of evidence and bodies of evidence. Hill wrote that they were just things to consider (that is, not criteria) when weighing evidence of causation. He used the five considerations from the Surgeon General's Committee plus four more to make these nine: specificity, strength, consistency, coherence, temporal relationship, biological gradient, plausibility, experiment, and analogy. Many causal assessors since then have used Hill's considerations, sometimes modifying them to fit their case and their concept of what constitutes weight of causal evidence. Currently, this approach is still the most common in human health assessments; ecological assessments use a greater variety of approaches.

Diagnostic procedures are a distinct type of weight of evidence. When fish, wildlife, plants, or humans are found dead, sick, or impaired, a pathologist with the appropriate expertise will use the procedures of his science to compile evidence and infer the cause. Examples of evidence include histopathology of gills, spots on leaves, identification of microbial pathogens, and organ concentrations of chemicals. A pathologist may diagnose the proximate cause by following a diagnostic manual or by case-specific weighing of the evidence. A pathology report may express sufficient confidence to define the proximate cause by itself. However, the assessment of causation can often benefit from other evidence such as a source for the proximate cause or spatial/temporal co-occurrence with an accidental release.

The current state of practice in weight of evidence for environmental assessments is methods that are structured by a process that defines what is being weighed and how. The methods respond to criticisms of weight of evidence by making it transparent and formal without losing the flexibility to analyze heterogeneous bodies of evidence. Some examples are EFSA (2017), Menzie et al. (1996), Australia and New Zealand (2017), ECHA (2010), and USEPA (2016e). The USEPA's weight-of-evidence guidelines are explained in the following chapter. These weight-of-evidence methods differ in their specifics. However, they all have in common that, unlike narratives or Hill's considerations, they have a well-defined and structured inferential process.

A variety of techniques have been labeled weight of evidence that have little in common with its use in practice. These include quantitative methods such as the log of Bayes factor, decision analysis, and Bayesian networks. These methods are not discussed here, because they stretch the already broad concept of weight of evidence to the breaking point. When lawmakers in the U.S., Europe, and elsewhere require that assessors weigh the evidence, I am sure that they did not mean any of those methods (USC 2016; Agerstrand and Beronium 2016).

Weight of evidence is the ultimate inferential tool. Most inferential tools function as methods to convert information into evidence concerning the assessment endpoint or an intermediate value. The resulting evidence that is derived by data analyses or models contributes to the overall weight of evidence. An example of the recognition of that role is in the web site for the AQUATOX aquatic ecosystem model; "Like any model, it is best used as one of several tools in a weight-of-evidence approach" (USEPA 2021b). Weight of evidence may be used as the inferential process for a complete assessment such as the USEPA's causal assessments (Chapter 11). It may also be used to derive a particular property or value in an assessment. For example, two recent publications provide weight-of-evidence approaches for determining whether a chemical is persistent (Redman et al. 2021; Hughes, Griffiths, and Swansborough 2021). In sum, if you do not weigh the body of evidence, you are leaving information on the table.

24.7 An Inferential Framework for Multiple Pieces of Evidence

In many assessment problems, more than one inferential method may be applied. In such cases, it may be apparent which method should be applied, but a framework for choosing an inferential approach can guide the choice and provide transparency in the process. An example is provided by a framework for making inferences from multiple pieces of evidence. The framework grew out of misunderstandings among assessors about the relationship

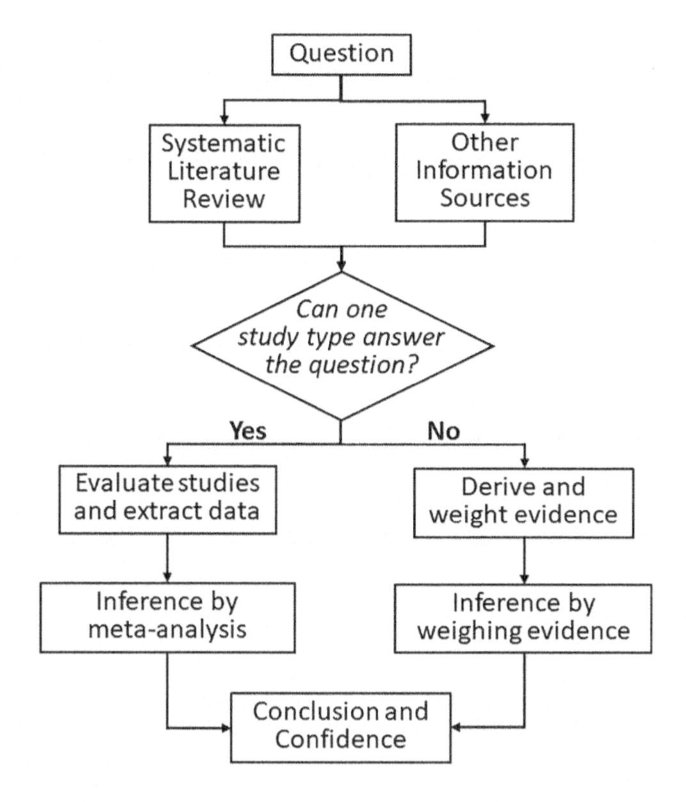

FIGURE 24.1
A framework for answering an assessment question when multiple pieces of evidence are relevant and must be combined by an appropriate inferential process. (Figure modified from Suter, Lavoie, et al. 2020b.)

between systematic reviews (Chapter 21), and the inference methods applied to their results. Some colleagues and I developed a framework to clarify those issues (Figure 24.1) (Suter, Lavoie, et al. 2020b). It does not include the full assessment process; it lacks planning, problem formulation, characterization, and communication (Chapter 3). Rather, it shows how to get to a conclusion concerning questions like is the chemical a carcinogen or teratogen (USEPA 2005b), does it biomagnify (Arnot et al. 2022), or is it the cause of a bird kill (Breinlinger et al. 2021).

The framework begins by identifying the question to be answered. It then obtains information that may provide evidence for answering the question by systematically reviewing the literature and, if possible, generating information to fill gaps or improve the body of the evidence. If the question can be answered by evidence from one type of quantitative study, studies of that type that addressed the agent or action of interest are assembled. The data from those equivalent studies is then combined by meta-analysis to generate an estimate and confidence interval. This is the approach of the Cochrane Collaborative which combines clinical trials of a drug or procedure to estimate effectiveness for a particular condition in humans (Higgins et al. 2021). If, as in most cases, multiple types of studies generate multiple relevant types of evidence, the evidence is weighted and the body of evidence is weighed (Chapter 25). If the meta-analysis, unlike those of clinical trials, does not include all high-quality information it is not convincing alone. In such cases, its results contribute to the weight of evidence. In any case, the result is a conclusion that includes a quality or quantity and an expression of confidence.

25

A Weight-of-Evidence Process

In our reasonings concerning matter of fact, there are all imaginable degrees of assurance, from the highest certainty to the lowest species of moral evidence.
A wise man, therefore, proportions his belief to the evidence.

—*David Hume (1748)*

I believe that weight of evidence is the most important inferential method in science, the law, and even day-to-day decisions (Chapter 25). However, until recently, it has not been taken as seriously as it should by environmental assessors. In this chapter, you will learn the basics of the USEPA Risk Assessment Forum's guidelines for weight of evidence (USEPA 2016e). They were developed by ecological assessors, but they were deliberately designed to be useful for human health assessments as well and were reviewed by human health assessors in the Risk Assessment Forum. For example, see an application of the method to derivation of human health benchmark values by read-across (Suter and Lizarraga 2022). With all due modesty, I believe that these guidelines are the best general weight-of-evidence approach available for environmental assessment.

25.1 Aspects of Evidence

When developing the USEPA approach, we put Hill's considerations aside, and we reconsidered the relationship between evidence and an inference. The result was five aspects of the evidence: aggregation, implication, weight, and properties. Each of Hill's considerations is an instance of one of the aspects. For example, Hill's consideration "strength" is one of the three major properties of evidence, and "experiment" is a type of evidence. Having defined those aspects, we could determine what Hill missed. For example, besides strength, relevance and reliability are major properties of evidence that should also be weighted, and there are many more types of evidence than experiment. The aspects are defined here before using them in the description of the weight-of-evidence process.

DOI: 10.1201/9781003156307-28

Aggregation of evidence. The degree of aggregation of the evidence determines how evidence will be assigned weights. In particular, each piece of evidence may be weighted, or evidence may be aggregated by types and each type weighted, and ultimately, the body of evidence is weighted.

Pieces are the essential units of evidence. Typically, a piece is evidence derived from a particular experimental or observational study. However, sometimes it takes information from multiple studies to derive a piece of evidence and sometimes a study provides multiple pieces of evidence.

Types of evidence are the categories that assessors naturally think of when describing evidence. Examples include results of rodent acute lethality tests, cell culture mutagenicity tests, fish tissue concentrations, and effluent concentrations. However, there can be different levels of aggregation of types, such as the very specific rat acute lethality tests versus the more general mammalian toxicity tests.

Bodies of evidence are all of the evidence that applies to particular hypotheses.

Implications of evidence express the logical relationships between evidence and inferences concerning the hypothesis. For example, laboratory toxicity tests may provide evidence of the mode of action of an agent, and surveys of biotic communities may provide evidence of co-occurrence of the effect and a potential cause. Without an implication, information is not evidence. (Implications were not sufficiently emphasized in the WoE guidelines. This became apparent as the guidelines were applied.)

Weight is an expression of the importance to an inference of a piece, type, or body of evidence. That is, the weight of evidence is the influence that the evidence has on your judgment concerning a hypothesis.

Properties of evidence are the aspects of a piece or type of evidence that determine how much weight it should have. This approach uses three general properties: relevance, strength, and reliability. These properties have subproperties which are appropriate to a type of inference.

Properties of bodies of evidence are collective properties of a body of evidence that provide weight including number, coherence, diversity, and signs of bias.

25.2 A Process for Inference by Weight of Evidence

Neither of the most common methods for weight of evidence in environmental assessment (narratives and Hill's considerations) has a defined inferential process (Chapter 24). However, structured weight-of-evidence processes are becoming more common. The USEPA's process is represented in Figure 25.1. Note that this is not a full assessment process (Figure 3.1), but rather an inferential component of an assessment. It does not include the impetus, planning, much of problem formulation and characterization steps, and communication of results. It may encompass most of an assessment, particularly a causal

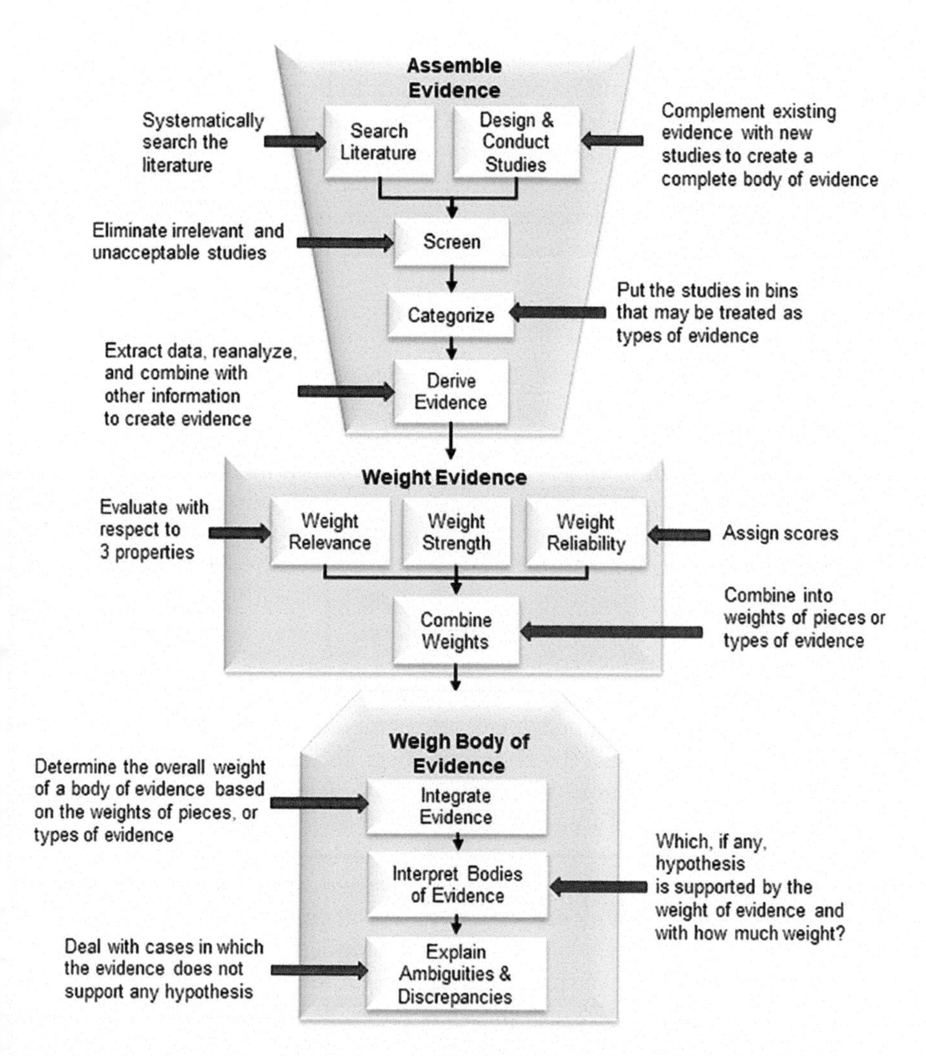

FIGURE 25.1
The process framework for the USEPA (2016e) guidelines for weight of evidence. Explanatory annotations are provided for each step.

assessment (Figure 3.3) or derivation of a benchmark value, or very little of an assessment, as in the derivation of a chemical property. The process is quite general and can be adapted to the specific needs of a program or individual assessment.

25.2.1 Assemble Evidence

The first step is assembling evidence, which begins by deriving new information or by finding existing information that can be used to produce evidence (Chapter 21). Assessment-specific studies should produce weighty evidence because it should be highly relevant to the assessment and should be more reliable than information from the literature. The greater reliability comes from your input as an assessor to specifying study designs and quality assurance and from your access to the original data.

Much or all of the evidence in most assessments comes from searching the literature or the web for useful information. During most of my career, this was done informally, but now it should be done in a systematic manner (Chapter 21). That includes at least identifying the sources searched and the search terms.

Both studies in the search results and studies performed for the assessment should be screened to eliminate irrelevant or clearly unreliable studies. This can be done in a separate screening step or during weighting by assigning a weight of 0. Elimination criteria should be defined in advance, to the extent practicable, to avoid potential bias. For example, in an assessment of risks to humans or mammalian wildlife, you might eliminate studies of invertebrate species as irrelevant. Reliability is less easily defined in advance. However, some criteria are obvious such as eliminating tests with no controls. Some organizations have standard screening criteria.

Studies are typically categorized into types. A basic 2×2 typology divides studies by design (experimental and observational) and location (laboratory and field). At a more detailed level, experimental toxicity studies may be divided into those that test the endpoint effect, those that test related whole-organism effects, and those that test for suborganismal mechanisms. The types into which studies are sorted depend on the assessment and the available studies.

The information extracted from the screened and categorized studies must be converted to evidence or its evidential properties must be identified. This requires identifying the potential implications so as to determine the form that the evidence should take. Convenient sets of implications are provided by the characteristics of causation (Chapter 11) and other qualities that you ascribe to the system being assessed (Appendix E of USEPA (2016e)). You may think of implications as equivalent to the "means, motive, and opportunity" that prosecutors seek to establish concerning a criminal defendant.

Evidence for a particular implication may require information from multiple studies. For example, if your hypothesis is that ozone caused observed foliar damage and your information includes a concentration–response relationship for ozone, that is only part of a piece of evidence. To have evidence of sufficient exposure (a causal characteristic), you also need an estimate of gaseous concentration at the time that the foliar damage occurred. Data reanalysis, including statistics, unit conversions, and normalizations, may also be required. For example, if the toxicity data is for ozone and the ambient concentrations are for total photochemical oxidants, a conversion is required.

25.2.2 Weight the Evidence

The heart of weight of evidence is assigning weights to the evidence.

How to categorize evidence? Categorization of evidence into types may be simply a means of keeping similar evidence together to keep the process organized. In that case, the individual pieces of evidence are organized into types before being individually weighted. However, it may be efficient and appropriate to assign aggregate weights to a type of evidence. This is a matter of practical judgment based on the abundance of evidence and the similarity of evidence within types. When evidence is complex, it may be helpful to categorize the evidence in terms of implications rather than types before assigning weights.

What properties? The three general properties of evidence (relevance, strength, reliability) may be evaluated (for example, how reliable is this piece of evidence?), but it is better to evaluate subproperties (for example, how well was the study designed, was the sample size sufficient, and has the study been replicated?) and then aggregate them.

Relevance—It has three recommended subproperties:

1. Biological relevance is the correspondence between the taxa, life stages, and processes measured or observed and the assessment endpoint.
2. Physical/chemical relevance is the correspondence between the chemical or physical agent tested or measured and the chemical or physical agent of concern.
3. Environmental relevance is the correspondence between test conditions and conditions at the assessed site or between the environmental conditions in a studied location and in the location of concern.

Strength—It is a property of the evidence obtained from the results of a study, not the reliability of the design of the study or its methods. A strong signal is better differentiated from noise than a weak signal, so a strong signal is given more weight. Most statistical analyses relate to strength, and it is easier to define standard scoring criteria for strength than for the other properties. Strong evidence shows

1. a large magnitude of difference between a treatment and control in an experiment or between exposed and reference conditions in an observational study,
2. a high degree of association between a putative cause and effect, or
3. a large number of elements in a category of evidence (e.g., multiple biomarkers give evidence of interaction).

Reliability—It consists of properties of a study that can make evidence trustworthy. There are many potential subproperties of reliability including design and execution, abundance of data, treatment of confounders, specificity, signs of bias, peer review, transparency, corroboration, consistency, and consilience. Assessors must consider which subproperties are most important to distinguishing the reliability of evidence in their case.

How to evaluate and score? The weighting of evidence includes evaluation and scoring. Evaluation determines the weight which is the influence a piece or category of evidence should have based on the defined properties. Scoring uses symbols to represent the results of the evaluation. The weight is conceptual, and the scores are symbolic. For example, a piece of evidence may be evaluated as moderately relevant, so the score for relevance is ++.

Evaluation involves defining your concept of weight for a property of evidence. For example, you might define ranges of correlation coefficients for strength of co-occurrence of a causal agent and an effect (e.g., an r^2 of 0.4 to 0.6 may be somewhat supporting for strength and would be scored +). You might also have rules such as, for human health assessment, human data are exceptionally relevant, mammalian data are strongly relevant, and other vertebrates are somewhat relevant. Examples of tables of rules for assigning weights can be found in USEPA (2000b) and Walls, Jones et al. (2015). However, most evidence evaluations are made without rules, and those judgments should be explained. The evaluation terms may vary among properties and evidence. For example, the occurrence of an effect prior to introduction of a source is convincingly negative evidence against that source as the cause. So is a large negative correlation between a toxicant and the magnitude of the effect of concern. The derivation of evidence from humans exposed to the contaminant source of interest makes the evidence for health effects exceptionally relevant. However, because of the difficulties with epidemiology, you would not call it convincing. Use terms that appropriately characterize the weight.

The scores, sometimes called ratings, are symbols that represent the weights. When tabulating and integrating weights, symbols are easier than words to scan for patterns. Symbols are preferable to numerical scores because they do not suggest that weights are quantities that can be added or subtracted. Two strongly supporting laboratory tests (++ & ++) are not equal to four somewhat supporting field tests (+ & + & + & +). For reliability of a test result, a – score for study design and a + score for replication of the test do not add to 0 because negative evidence is more convincing (see Carl Popper). The scores signify different levels of qualitative properties.

The following is a set of terms for scoring weights with corresponding symbols:

Convincingly supports or weakens	+++, – – –
Strongly supports or weakens	+ +, – –
Somewhat supports or weakens	+, –
No effect (neutral or ambiguous)	0
No evidence	NE
Not applicable	NA

TABLE 25.1

Partial WoE Table for Alternative Possible Causes of the Decline of San Joaquin Kit Foxes

Types of Evidence	Predation	Toxics	Accidents	Disease
Spatial/temporal co-occurrence	+	+	+	–
Temporal sequence	0	0	NE	NE
Evidence of exposure or biological mechanism	++	++	++	––
Causal pathway	+	+	+	0
Manipulation of exposure	+	NE	NE	NE
Stressor–response relationships from simulation models	+++	–	+	––
Coherence	+++	–	+	+

Source: From Suter and O'Farrell (2015) and USEPA (2016e).
For illustrative purposes, only 4 of 6 potential causes and 7 of 16 types of evidence are included.

25.2.2.1 How to Tabulate Weighted Evidence?

The primary tool of this approach is an evidence scoring table. A table for a hypothesis should present the properties and scores for each piece or category or evidence. The table may also include the justifications for the scores, or they may be provided in the text or footnotes.

25.2.2.2 How to Combine Weighted Evidence?

After the properties of evidence are evaluated and scored, it is often helpful to combine the weights across properties to derive an overall weight for each piece or category of evidence that you have scored. The result is a combined weight (Table 25.1). Combine scores logically rather than adding the scores. For example, a study may be highly relevant (+++), but if the correlation coefficient for exposure and response is negative (– – – for strength) or if it does not have an appropriate control treatment (– – for reliability), the combined score should be negative. In other words, the high relevance of the study gives you greater confidence in the evidence against the hypothesis. If there is relatively little evidence, you may skip this step and carry forward the evidence weights for all of the properties to the weighing of the body of evidence.

25.2.3 Weigh the Body of Evidence

Weighing the body of evidence is the crucial step in which the overall weights of evidence for each hypothesis are inferred and interpreted. If all goes well, you will identify a best and sufficiently supported hypothesis.

Your first step is to integrate two types of evidence: (1) the weighted individual pieces or types of evidence, as described above, and (2) the weighted collective properties of the body of evidence. Collective properties of the body of evidence may include number of pieces of evidence, their coherence and diversity, and signs of bias. Number may be number of studies or number of categories of evidence for each hypothesis. Coherence is the degree to which the body of evidence makes sense. Diversity is the range of evidence assessed such as taxa, life stages, or habitat types. Signs of bias are differences in the results of studies that appear to be related to employment, funding, or other conflicts of interest. When the collective properties have been assigned weights and associated scores; they can be combined with the weights for the evidence from the previous step in a weight of evidence (Table 25.1). The overall weight of evidence for a hypothesis is an expression of confidence in that hypothesis. As discussed previously, this weighing is not a matter of adding the scores.

Interpretation is the process by which you make sense of the weights of the bodies of evidence for the alternative hypotheses. In some cases, this is simple. If all hypotheses have sufficient evidence to be evaluated, and one hypothesis has clearly more weight than the others across categories and properties of evidence, you accept that hypothesis. In more difficult cases, your interpretation may be aided by considering the implications of the evidence. For example, in causal assessments, it may be helpful to consider how many of the characteristics of causation (a set of implications for causal hypotheses) are supported by evidence. Evidence for a hypothesis may have predominately positive weights, but it might all be associated with few characteristics, and they are indicative only of antecedence and co-occurrence (Chapter 11). If you have made conceptual models for the hypotheses, interpretation may be aided by relating the weighted evidence to links in those models.

In some cases, you must explain ambiguities and discrepancies. It may be that all hypotheses have incoherent evidence which was not resolved during interpretation, or none have predominately positive evidence. It may be that more than one hypothesis has predominately positive evidence. For example, in the causal assessment of the San Joaquin kit fox decline, predation by coyotes had abundant and weighty positive evidence (Suter and O'Farrell 2009, Suter and O'Farrell 2015). Loss of prey due to drought also had positive evidence but the foxes were not starving or even lean. The ambiguity was resolved by redefining the loss of prey hypothesis as causing the foxes to spend more time hunting, which made them more susceptible to predation. Ambiguities and discrepancies may be due to poorly specified hypotheses. For example, studies of the cause of peregrine falcon decline were unsuccessful until the endpoint was redefined more specifically as reproductive failure associated with eggshell thinning. Ambiguities and discrepancies are often due to differences among hypotheses in the amount of evidence and particularly in their distribution among types or implications. In such cases, you may conclude the assessment by calling for new and better targeted studies.

25.3 Weight of Evidence for Quantitative Results

After evidence is weighed to assess a quality such as causation, teratogenicity, or bioaccumulation, you may follow up by deriving a quantity such as allowable concentration, threshold value, or bioaccumulation factor. If the available information provides more than one estimate of the quantity, you may employ either of two approaches to derive a value (USEPA 2016e; Suter, Cormier et al. 2017). If the sources all yield the same type of result, such as all are threshold concentrations for deformities in fish, meta-analysis may be used to combine them (Figure 24.1). If they are not the same type, weight of evidence may be used to identify the best (weightiest) value. You may adopt either of those as the quantitative result.

25.4 Practicality

A weight-of-evidence analysis may be complex and time consuming. This is justified by complex and important assessments such as the kit fox decline and read-across for DDD (Suter and O'Farrell 2009; Suter and Lizarraga 2022). However, if the hypothesis to be analyzed is simple and the body of evidence is small, it is reasonable to reduce the body of evidence to one or a few types or implications of evidence and to weight a few influential properties of the evidence. In any case, it is important that you perform a sufficient analysis to assure yourself and others that the relevant evidence has been objectively weighed and the conclusion is defensible.

26

Extrapolation

Ideally, you will have data that represent your endpoint entities and attributes and that are generated by studies in conditions that are representative of the environment being assessed. More often, you must estimate the quality or quantity of interest by extrapolating from available but substantially different data. Extrapolations are made between similar entities such as from a measurement of the sensitivity of a tested fish species to an untested fish species. Extrapolations extend the applicability of information. Estimating a property of an entity from a very dissimilar entity is not extrapolation because it is a change of type. For example, using a chemical's structure to estimate the sensitivity of fish (i.e., a Quantitative Structure–Activity Relationship) is simply empirical modeling (Chapter 24). Extrapolation can be performed by analogy, application of a factor, or statistical analysis.

26.1 Analogy

As discussed in Chapter 24, analogy is a process of estimating an attribute of an entity by assuming that it will be the same as that of a similar entity. Analogy is conceptually the simplest extrapolation method. That is, we can simply assume that attributes of an entity can be extrapolated to sufficiently similar entities.

The most common extrapolation by analogy is the extrapolation of effects and sensitivity to the effects from tested species to untested species. For example, by analogy, we assume that taxonomically similar species will respond similarly to an agent. Hence, when assessing risks from aqueous exposures to chemicals, it is commonly assumed that a sufficient base set of tests is one fish, one invertebrate, and one alga (EC 2011). Freshwater fish are expected to respond similarly (with some error due to variance), so one fish species such as fathead minnow or zebra fish is often sufficient to extrapolate to any other freshwater fish species. If you are assessing risks to a specific untested species from tests of multiple fish species, you may extrapolate from the data for the most taxonomically similar species. For example, when assessing risks of copper toxicity to Bristol Bay salmon, I was confident in the analogy from rainbow trout because that test species is from the same genus as all of the salmon species (*Oncorhynchus*).

DOI: 10.1201/9781003156307-29

Similar analogies are made among ecosystems. By analogy, you might use data from one Appalachian stream to extrapolate to another stream but not from a stream to a pond or even from an Appalachian stream to a coastal stream. More similarity of the ecosystems imparts greater confidence to the extrapolation. Although there is no generally applicable taxonomy of ecosystems, you can assign similarity by judgment guided by defined criteria.

You can also extrapolate properties between chemicals and between other pairs of similar agents by analogy. If the agent for which you are deriving a benchmark has not been tested for the endpoint of interest, you may use a process called read-across to identify the best adequately similar analog that has been tested (the source chemical). Then you extrapolate from the tested source chemical to an untested target chemical. That is, the benchmark value for an inadequately tested "target chemical" is the benchmark value for its analog (Chapter 23).

Most reviews of extrapolation do not discuss analogy as a method. Rather, it is simply treated as an assumption. However, as simple as it is, analogy is a legitimate and common extrapolation method, and it deserves more careful consideration. In particular, although taxonomy is the best criterion for judging similarity of species, it may be useful to include other properties of species, such as size, diet, or morphology, in a weight-of-evidence method similar to structured read-across methods.

26.2 Factors

The most common extrapolation method is dividing an available value by a factor. The factor may represent an extrapolation between a test endpoint such as a *Daphnia* EC_{10} and an assessment endpoint such as invertebrate species richness. An extrapolation factor of 3 might be applied, because experience suggests that some invertebrate taxa is likely to be more sensitive than *Daphnia*, but not a lot more. A factor may also represent uncertainty. In the same example, you may have no evidence concerning the toxicity of the chemical to algae, but you do know that *Daphnia* responses have different toxic modes of action in most cases. Since the *Daphnia* response is likely to be qualitatively different from algae, an uncertainty factor of 100 may be applied for lack of knowledge. In most cases, factors represent a combination of extrapolation and uncertainty and are generally referred to as assessment factors. Assessment factors are intended to protect the assessment endpoint with reasonable confidence. For example, see the USEPA assessment factors for extrapolating from various sources of aquatic toxicity data to aquatic communities when assessing risks from non-pesticides (Table 26.1).

Most factors are based on expert judgment and some statistical analyses concerning what value would be adequately protective. As a result, most

TABLE 26.1

Factors for Extrapolation in Risk Assessments for Non-Pesticide Chemicals (Nabholz, Clements et al. 1997)

Toxicity Data	Factor
Quantitative Structure–Activity Relationship	1,000
Acute data (1 or 2 taxa)	1,000
Acute data (3 taxa)	100
Chronic data (3 taxa)	10
Meso- or microcosms	1

Each type of toxicity data is divided by its factor to estimate an acceptable level of exposure.

extrapolation factors used in human health and ecological assessments are factors of 10 because there is no basis for better precision.

Conventional factors for human health assessments in the U.S. are factors of 10 for animal to human extrapolation and for typical human to sensitive human extrapolation. Another factor of 10 may be used for uncertainty due to lack of information. A partial factor may be applied, usually when missing information is not so consequential. That partial factor is the square root of 10, rounded to 3. The factors are multiplied. For example, the factors for extrapolation from a single rat test to typical human to sensitive human with incomplete data could be $10 \times 10 \times 3 = 300$. Other factors may be used in special cases. Factors are the standard extrapolation method for human health assessments.

26.3 Statistical Extrapolation

Some factors are derived from simple statistics. For example, when deriving the method for calculating acute ambient water quality criteria, USEPA scientists decided that, even though the standard test endpoint is the LC_{50}, the death of half of the exposed organisms in a species was unacceptable. To extrapolate to a lower death rate, they collected acute aquatic test data and determined the mean of LC_{50}/LC_{10}, to one significant figure, was 2. That factor of 2 has been used when calculating acute water quality criteria to protect 90% of organisms from acute lethality since 1985.

Other extrapolations use regression models to relate effects in one taxon to those in another. The USEPA's method, Interspecies Correlation Estimation (ICE), regresses one taxon (species, genus, or family) against another (Raimondo, Lilavois et al. 2016). After decades of development, it has begun to be used to estimate toxicity to an untested species of interest. It also provides additional species for species sensitivity distributions for chemicals that have been tested in few species. See for example, Attachment 4-3 of the assessment of risks to endangered species from chlorpyrifos (USEPA 2021c).

26.4 Levels of Organization

In some cases, you can extrapolate to population and community qualitative effects from organism test results based on knowledge of how the tested organisms represent or contribute to the higher level system. For example, applications of insecticides at doses intended to kill arthropod pests will kill nontarget arthropods. That mass mortality will reduce arthropod population and assemblage abundances, which may result in starvation of insectivores and even predators of insectivores. That chain of effects was documented for neonicotinoid insecticides in a Japanese lake (Jensen 2019; Yamamuro, Komuro et al. 2019). As the authors state, it takes years of field studies to prove that the observed loss of fish is a food-web-mediated effect of a pesticide, including studies to eliminate alternative causes. However, if the use of neonicotinoids on rice results in exposure to concentrations that are lethal to aquatic insects (an organism-level effect on an ecologically important taxon), you can predict that some adverse population and ecosystem effects will occur. That qualitative extrapolation is enough in many environmental management contexts.

In some other cases, your reasonable expectations concerning higher level effects may not be confirmed by field studies. For example, the reported population-level effects on pelagic fish of the huge 2010 Macondo oil well blowout are less than expected given the results of organism-level tests and even from observed effects of prior oil spills (Fodrie, Able et al. 2014). These discrepancies may be due to inadequacies of the tests, the field studies, or both. Some fisheries were significantly diminished, and those effects would conventionally be considered to constitute unacceptable effects in most assessment contexts. Nevertheless, the requirement for natural resource damage assessment in the U.S. Oil Pollution Act meant that magnitudes of effects must be estimated. Those estimates depended on field studies and fisheries data, not extrapolation from laboratory tests. This is a reminder that you need to know your legal and regulatory context.

Models to extrapolate organismal effects to population effects are numerous and useful in many cases. A particularly interesting example is the linking of two standard avian models, TIM for acute lethality and MCnest for reproductive success, to estimate population effects in 31 avian species of applications of 12 insecticides (Etterson, Garber et al. 2017). The interesting finding was that the magnitude of effects was due largely to the timing of applications relative to the timing of breeding activities. That is, an exposure trait was most important for extrapolation among species of birds.

27

Species Sensitivity Distributions

Species sensitivity distributions (SSDs) are exposure–response models for biotic communities. You may wonder why, of all types of exposure–response models, SSDs are worthy of their own chapter. First, they are particularly important because, since 1985, they have been the standard tool in derivation of water quality benchmark values in the U.S. and many other nations. Second, they are often misunderstood. Third, their derivation requires many decisions relative to other exposure–response models. Fourth, I have a personal attachment to SSDs. I identified them as a distinct type of model and gave them their name at an OECD (1992) workshop. Then, SSDs gave me the opportunity to work with excellent scientists at the Rijksinstituut voor Volksgezondheid en Milieu (RIVM) in reviewing the uses of SSDs and elucidating issues in their implementation (Posthuma, Suter, and Traas 2002).

I would like to explain what SSDs are and are not. They are often referred to as extrapolations (including by me at one time), but in their form and use, they are exposure–response models. SSDs are models of the distribution of the proportion of species affected relative to exposure levels (Figure 27.1). They are like the exposure–response models used to derive LC_{50}s or EC_{10}s, which are distributions of organisms affected relative to exposure levels (Chapter 7). SSDs are models of communities in terms of the sensitivities of their constituent species just as the organism-level exposure–response relationships are models of populations in terms of the sensitivities of their constituent organisms. Also, although SSD models estimate proportions of species affected, some practitioners state that they estimate probabilities, without specifying probabilities of what (Chapter 18) (Suter 1998). Truly probabilistic SSDs go beyond estimating proportions and derive probability distributions of the proportions based on variance of a fitted model, resampling, or even treating each species value as a distribution (Wigger et al. 2020).

The most common use of SSDs is to derive benchmark values (Chapter 22). That use involves deriving the SSD and solving for a proportion of species affected to obtain a benchmark concentration. The estimate of that proportion, usually 5%, is called the hazardous concentration for that centile (HC_5). Less commonly, SSDs are used in risk assessments to estimate a level of effects. That use employs the same relationship, but it is solved for an estimated exposure level to obtain an estimated effects level defined as a proportion of species affected (sometimes called the Potentially Affected Fraction, PAF). SSDs may also be used in causal assessments. That is, the observed

DOI: 10.1201/9781003156307-30

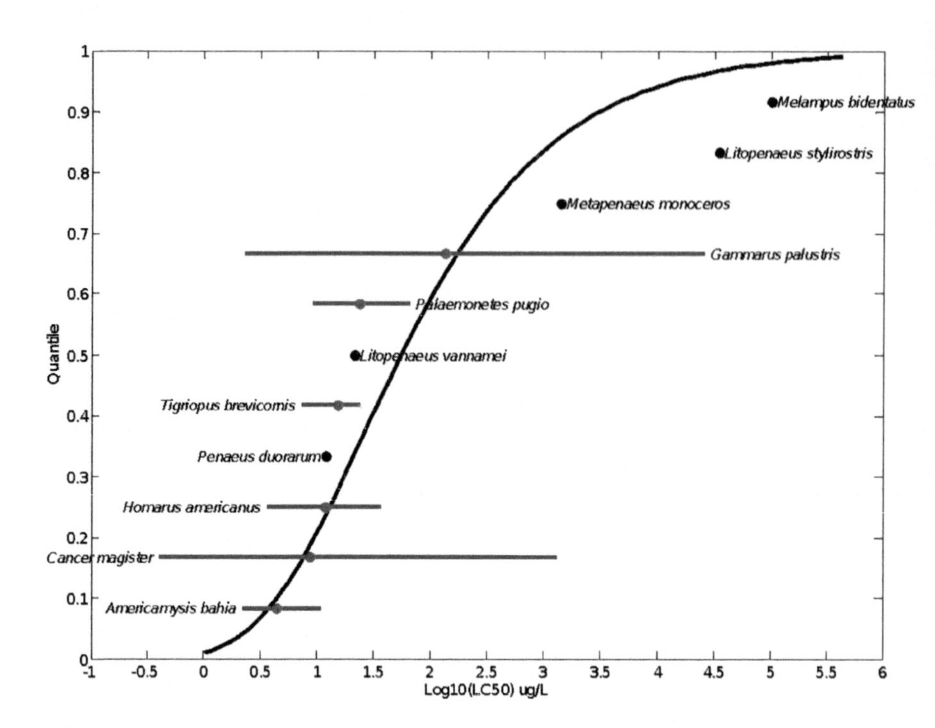

FIGURE 27.1
An SSD from a log-Gumbel distribution fit to malathion saltwater invertebrate data. Black points indicate single toxicity values. Red points indicate an average of multiple toxicity values for a single species. Blue lines indicate full range of toxicity values for a given species. From USEPA (2022a).

relationship of concentrations of a potential cause to the magnitude of community effects in the field can be compared to an SSD derived from results of laboratory tests of the potential cause. Similar laboratory- and field-based community exposure–response relationships constitute evidence in support of the causal hypothesis.

If you want to derive an SSD, you have two types of implementation decisions to make: choosing the method for deriving the results and choosing the data to include. Four approaches have been used to derive SSD results. First, you can fit a parametric statistical distribution to the data, most often the normal or log-normal (Etterson 2020; Aldenberg and Jaworska 2000). This parametric approach can be problematical because of the frequency of poor fits of the function to the data (Newman et al. 2002). Second, Newman's solution to the problem is to use the bootstrapped lower 95% confidence bound on a nonparametric estimate of the HC_5 as the benchmark value. This approach is less susceptible to irregular data distributions and is likely to be protective, but unconventional. Third, the form of the distribution is often different for the sensitive and resistant species (lower end versus higher end) resulting in asymmetric distributions. When deriving benchmarks, we are concerned only with the lower end of the distribution. Therefore, USEPA's solution is to fit a log triangular distribution to the lower end data (Stephan et al. 1985). Fourth, greater than and less than values cannot be fit, although if there are few of them, they can be estimated by assuming a distribution. Data sets that lack values for lower and upper ends are referred to as censored and estimating them is uncensoring. Censored data cannot be ignored because they influence the distribution by contributing to the proportions. In particular, SSDs derived from field data may include many species that are resistant to all contaminant concentrations in the sampled water bodies. Further, field data like lab data often do not have a nice symmetric sigmoid distribution (Figure 27.2). A simple solution to these problems is linear interpolation which requires only that the values in the vicinity of the benchmark value be defined (Cormier, Suter, and Zheng 2013; USEPA 2011b).

The second problem that you must address is, what taxa and how many taxa should be included? One the one hand, you might include all taxa for

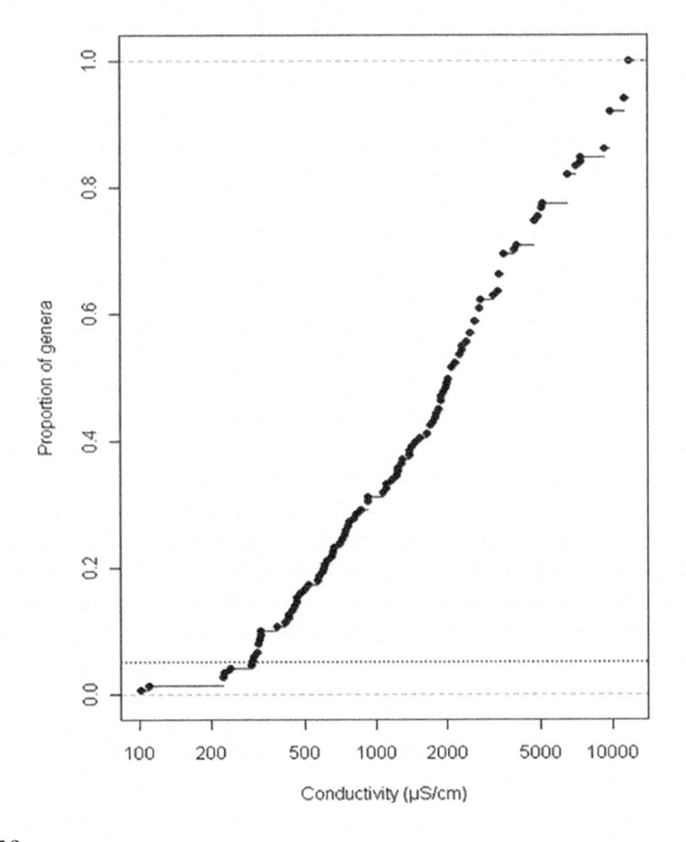

FIGURE 27.2
An SSD for the proportion of stream insect genera extirpated by major ions, measured as specific conductivity, in West Virginia streams (USEPA 2011b). It illustrates the large number of taxa available in the field to derive an SSD even though insects are identified only to genus.

which you have data, so that the SSD might represent the entire community. On the other hand, you may wish to disaggregate taxa to estimate more specific risks or taxa-specific benchmarks. For example, when deriving water quality criteria, the USEPA combines vertebrates and invertebrates. However, it could be more informative to a decision-maker to separately estimate the proportion of vertebrates and invertebrates affected, particularly for pesticides. The USEPA does not combine plants and animals in an SSD when deriving criteria, but European assessors routinely combine them. Finally, as with all empirical modeling, you must decide how many data (species) is enough. SSDs for USEPA criteria require eight species and they must cover prescribed taxa. Some European organizations require a minimum of three species but add a safety factor. In sum, you need to know if there are policies or standard practices in your organization for deriving SSDs for benchmarks or risk assessments.

A third problem is to determine what quality constraints should be imposed. Should you include nonresident or nonnative species, nonstandard test designs, extreme water chemistries, short durations, or other quality issues? The number of points in an SSD may be increased by using data derived from QSARs, read-across, taxonomic extrapolation models, or other methods. However, these techniques also reduce confidence in SSDs. Quality requirements may be loosened if uncertainty factors are applied to SSD results or if quality scores are applied to SSDs that determine their use (e.g., screening or regulation). A Dutch group developed a database of 12,386 SSDs with quality scores (Posthuma et al. 2019).

Your fundamental question is, should I use SSDs to represent community responses? Despite the popularity of SSDs for regulation, some assessors say no. One objection is that SSDs are biased because test species must be resistant if they survive in laboratories. However, it is clear that tolerance of laboratory conditions does not predict sensitivity to the various mechanisms of toxic action. Also, SSDs may be biased because untested effects are likely to be more sensitive than tested ones. That is plausible, but it is true of all assessment methods that are based on standard tests of a few population-relevant effects. These objections could be resolved by deriving field SSDs which integrate all effects that contribute to occurrence and abundance of species in real communities (Cormier, Suter, and Zheng 2013; USEPA 2011b; Leung et al. 2005; Schipper et al. 2014; Kwok et al. 2008). However, deriving SSDs from field data may be difficult in terms of obtaining and analyzing the data, dealing with confounding factors, and addressing site or region specificity.

Because of the importance of SSDs, the literature on SSD practices is abundant and growing. Recent papers such as Belanger and Carr (2020) and Fox et al. (2021) address all of the issues raised in this chapter and have come to various conclusions. There are several good SSD software tools. The USEPA's is *Toolbox* (Etterson 2020).

28

Mixtures, Multiple Agents, and Cumulative Effects

Fillet of a fenny snake, In the cauldron boil and bake; Eye of newt and toe of frog, Wool of bat and tongue of dog, Adder's fork and blind-worm's sting, Lizard's leg and owlet's wing, For a charm of powerful trouble, Like a hell-broth boil and bubble.

—*Shakespeare,* Macbeth

You will find that many assessments deal with a single agent such as a new chemical or an introduced species. However, in the real world, you encounter agents with multiple components, media containing multiple agents, multiple effects of an agent, and multiple effects of multiple agents. Terminology is inconsistent. We define an "undesigned mixture" as multiple chemicals and other agents that are not deliberately formed and vary in space and time. An example is the chemicals and suspended sediment in a stream or the accumulation of chemicals added to an agricultural soil. We distinguish "designed mixtures" as deliberately formed combinations of chemicals such as a pesticide formulation or a laundry detergent. An effect of a mixture of agents is referred to as their combined effect. The USEPA refers to the estimation of combined risks from multiple agents by multiple routes from multiple sources as cumulative risk assessment. The combination of multiple effects is referred to as cumulative effects.

28.1 Mixtures

Methods for assessing mixtures may be based on testing of the whole mixture or on testing its components and estimating their combined effects (Figure 28.1).

28.1.1 Whole Mixture Testing

The most straight-forward approach is to test the toxicity of the mixture of concern. Designed mixtures may be tested using standard single chemical test protocols. That is, a mixture such as a herbicide formulation may be added to a test medium as in single chemical tests (e.g., mg of mixture per

DOI: 10.1201/9781003156307-31

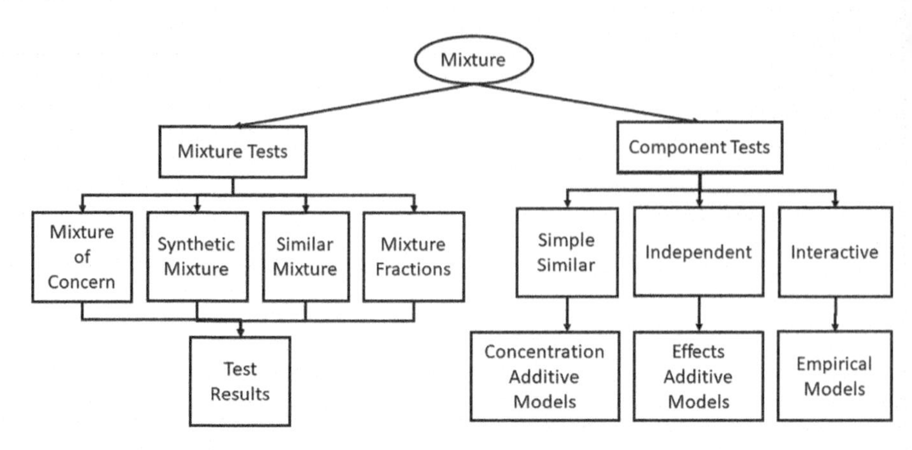

FIGURE 28.1
A diagram of alternative approaches to determining the effects of mixtures.

liter of water). Undesigned mixtures of concern such as effluents, highway runoff, and contaminated media may also be brought to the laboratory and tested. If it is toxic, a dilution series, preferably with uncontaminated local media, should be tested. If conditions allow, contaminated media along a pollution gradient may be collected and tested. The most prominent example of tests for effluents and contaminated media is the acute and 7-day semi-chronic tests developed for effluents and ambient waters by the USEPA (2002a, f, g). The semi-chronic tests are reasonably protective and practical for routine testing in fresh and salt waters.

The mixture may be simulated by synthesizing it from constituents. That creates a consistent and well-specified mixture to be tested. Alternatively, a mixture that is unavailable because its source does not yet exist may be simulated by synthesizing and testing the expected mixture.

Sometimes, mixtures are unavailable, but testing of a similar mixture may be an option. For example, when planning a new factory, effluent from an existing factory of the same type may be tested undiluted or diluted in water from the proposed site. Some criteria should be established in advance to determine whether the tested mixture and the mixture of concern are sufficiently similar (Catlin et al. 2018).

On the other hand, if a mixture is similar in different cases and the effects are sufficiently consistent, it may be useful to aggregate its constituents into a single exposure metric. For example, the mixture of major ions in fresh water are most easily measured as specific conductivity, the ionic activity of the mixture. The effects on stream communities of ion mixtures measured as specific conductivity are similar across surface-mining-influenced central Appalachian streams, despite some variance in composition. Similarly, total petroleum hydrocarbons is a common metric for analyses of petroleum spills due to the complexity of petroleum.

Fractionating a mixture may increase the likelihood of identifying specific effects and relating those effects to exposures. For example, when total petroleum hydrocarbons are found to be a significant hazard, it can be useful to fractionate the oil. The relative contribution of aromatic and aliphatic hydrocarbons and of high, moderate, and low molecular weight fractions can influence the fate and effects of a spill.

Testing of effluents or contaminated media for human health assessments is rare. This is because unlike fish and other nonhuman species, humans do not inhabit and seldom consume contaminated ambient media. For example, in most of the world, people seldom consume contaminated and untreated surface water. Therefore, mammalian testing of effluents or contaminated media for human health assessments is seldom relevant. Humans do routinely inhale contaminated air, but mammalian testing of air or atmospheric effluents is also uncommon for practical reasons. Mammalian testing of mixtures is largely limited to designed mixtures.

An alternative to testing contaminated ambient media is observational field studies of contaminated sites. Those epidemiological and ecoepidemiological studies are discussed in Chapter 21.

28.1.2 Components Testing

Rather than testing a mixture, you may estimate its toxicity using the toxicity of the components and assumptions concerning their combined effects. Three alternative assumptions are that the components all have similar action, they have independent action, or they are interactive. The analytical approaches for these three alternatives are models of (1) concentration or dose addition, (2) effects addition, and (3) synergism or antagonism. The type of combined action may be known from mechanistic studies of the constituent chemicals, but that is uncommon. Sometimes signs of a chemical's mechanism are supplied by observations of responses during toxicity tests or from in vitro screening tests. Sometimes mixtures tests are designed to reveal the mode of combined effects. More often, chemicals are assumed to have similar action if they belong to a common chemical class, have parallel exposure–response relationships, or are fit by a common Quantitative Structure–Activity Relationship (QSAR). Multiple techniques may be used to estimate effects for each assumption, but in the interest of simplicity, the equations and graphs needed to explain them are not included here. They are included in guidance documents from some agencies (Risk Assessment Forum 2000).

Similarly acting chemicals have the same mode of action and their toxicokinetics are qualitatively the same. Therefore, their exposure concentrations or doses can be added after normalization for differences in potency. If equivalent exposure–response relationships are available for all members of the similar mixture, they should be parallel (i.e., equal slopes), and the distances between the lines represent their relative potencies.

Therefore, their concentrations can be added when potency factors relative to one chemical are applied to each of the others. In that way, effects can be estimated for a mixture of similarly acting chemicals from the relative potency-weighted sums of concentrations. This technique may also be applied to species sensitivity distributions (Traas et al. 2002). If good equivalent exposure–response relationships are not available for all chemicals in a mixture (often the case), other measures of relative potency can be used (USEPA 2003a). For example, if exposure–response relationships for cancer are not available for all similar chemicals in a mixture, their relative potency as mutagens might be used to normalize those chemicals to an index chemical that has a high-quality exposure–response relationship for cancer. A simpler approach is to normalize the concentrations by dividing them by a common benchmark value or a common test endpoint such as a *Daphnia* LC_{50}. These toxicity-normalized concentrations are called toxic units (TU), and their sum is called a hazard index (HI). If the HI is greater than 1 and the chemicals have additive toxicity, more than half of *Daphnia* should die if exposed to the mixture. Concentration addition is a common default method for assessing mixtures.

Independently acting chemicals have the same effects, but do not have the same mechanisms and do not interact with one another. Hence, their effects can be added. If, at the expected exposure to the mixture, one chemical kills 5% of *Daphnia* and a second chemical that is independently acting kills 10%, 15% are expected to be killed by the mixture. Actually, because of the possibility of a daphnid receiving a lethal exposure from both chemicals, the probability of intersection (i.e., 0.5% in the example) should be subtracted so 14.5% are expected to be killed.

Chemicals in a mixture may interact in a variety of ways. Chemical interactions are most often antagonistic. Antagonism may occur when one chemical inhibits the uptake of another, they compete for binding sites in cells, or one promotes metabolism of another. Synergistic interactions can occur by the opposite processes such as one chemical inhibiting metabolism of another. Interactions can be quite complex including different interactions at different concentrations. Ideally, the interactions are simulated using toxicokinetic and toxicodynamic models. This requires a lot of work. Alternatively, you can test mixtures of chemicals at varying relative concentrations and empirically model the interactions. This also requires a lot of work. In general, the best approach to assess interacting chemicals is to test the mixture of concern by collecting or synthesizing the mixture. Synergistic effects are a concern because they suggest the possibility of toxic effects that were not expected from individual toxicities. However, many reviews over the years have found synergism to be rare, and even when it is detected, the effect is rarely more than a factor of 10 (Cedergreen 2014). However, some generalizations are useful. For example, some chemicals such as cholinesterase inhibitors and azole fungicides inhibit metabolism of other synthetic organic chemicals.

These models of combined effects are problematical when the constituents are numerous, and the mechanisms are not well known. Some mixtures contain hundreds of chemicals many of which are unknown and uncharacterized. One solution suggested by the European Chemical Industries Council is the lead component identification methodology (LCID) which bases the assessment of a mixture on one or a few most hazardous chemicals (Galert and Hassold 2021). Reducing the number of chemicals simplifies the assessment task but adds more assumptions to the process. Some directions for development of methods for assessing highly complex mixtures have been suggested (Chibwe et al. 2017).

At the other extreme, if you have good data for the physiochemistry of the members of a mixture, you may use toxicokinetic models to mathematically simulate any interactions in tissue dosimetry (USEPA 2007b, c). However, to apply this approach, you must also know the mechanism of action of the chemicals in the mixture (i.e., its toxicodynamics). This is likely to be true for pesticide mixtures, but not most others.

28.1.3 Multiple Types of Evidence

There are many approaches and specific techniques for assessing mixtures, and each one provides different evidence concerning the nature and magnitude of effects. The USEPA mixtures guidelines state, "all possible assessment paths should be performed" (Risk Assessment Forum 2000). So, this is another case in which a weight-of-evidence approach is desirable (Chapters 24, 25). Testing the mixture is generally better than combining testing results for the constituent chemicals. However, the mixtures test may be given low weight because it was poorly performed or performed with a minimally relevant species, life stage, duration, endpoint effect, etc. Tests of constituents also may have good or bad properties. Assumptions concerning additivity or interactions may be well or poorly supported. The weight-of-evidence process may determine whether the body of evidence supports a particular combined effect of a mixture and the best estimate of the magnitude of effects given exposure to the mixture.

The effectiveness of the weight-of-evidence process depends on the evidence that is possible for you to obtain or generate. If you cannot perform tests, you are stuck with using whatever tests of constituents are available and whatever assumptions are reasonable. This is impractical if the mixture is highly complex and particularly if its constituents are unknown (Lai et al. 2022). Ideally, you test the mixture of concern, but similar mixtures may also be useful if available. For either constituent or mixtures, you should be aware of the variability of mixture compositions. The most realistic approach is to monitor the response of organisms, populations, or communities exposed to the mixture. However, that means that people or nonhuman organisms must be exposed to the potentially toxic mixture. In addition, the problem of confounding may be insurmountable (Chapter 11). All of these issues would go into the weighting of evidence.

28.1.4 Mixtures Exposure

The estimates of exposure to a mixture must be concordant with the exposure metric in the exposure–response relationship. If you are estimating toxicity from tests of the components, you must measure or monitor the concentrations of those same components. That is, you analyze the concentrations of the individual chemicals in the test and in the field. If the concentrations are variable, it is good to match the averaging time of samples to the acute or chronic duration of the endpoint.

Exposure assessment for whole mixtures can be expressed as proportional dilutions rather than concentrations. For effluents, these may be undiluted, diluted at the edge of the zone of initial dilution, diluted when fully mixed, or diluted at a set distance below the discharge. The zone of initial dilution (mixing zone) is set at a default of 10:1 dilution in the U.S., but there are numerous case-specific considerations in both the U.S. and Canada such as avoiding partial dilution in sensitive areas and human recreation areas and allowing a zone of passage for migrating fish (Environmental Protection Division 2019). The various dilutions can be calculated if the effluent does not yet exist, but if contaminated ambient media are tested, they are already diluted. The exposure mixture in contaminated water, sediment, or soil should be the same as the tested mixture at same the time and place (Chapter 6). Just split samples for testing and analysis.

28.2 Cumulative Risk

The mixtures assessment techniques discussed above are designed to deal with a particular designed or undesigned mixture and are nearly always concerned with chemicals. Cumulative risk assessment goes a step further. It is designed to assess all of the agents acting upon an organism, population, or community in some context. It includes multiple chemical and physical mixtures, habitat quality, harvesting, pathogens, and severe events such as storms and floods (USEPA 2003b; Moretto et al. 2017; USEPA 2007b). In human health assessments, it includes psychosocial factors as well. There is no legal mandate for cumulative risk assessment in the U.S. However, it is sometimes used, particularly with respect to human health, as a broad causal assessment. For example, what are the causes of rates of cardiovascular disease in human populations (Kashuba, Menzie, and Martin 2021)?

28.3 Cumulative Effects

Cumulative effects are determined by combining the separate effects of multiple agents, rather than combining the agents to estimate their combined effects, as discussed above. In the U.S., the term cumulative effects is important because their consideration is mandated for impact assessments under NEPA (Chapter 1). In particular, the regulations require that cumulative effects include not only the effects of the action under consideration but also "the incremental impact of the action when added to other past, present, and reasonably foreseeable future actions regardless of what agency (Federal or nonFederal) or person undertakes such other actions" (CEQ 1997). For example, an environmental assessment of widening a road must consider the cumulative effects of the habitat lost due to the widened road and not just the incremental loss from the new lanes. A classic example is the National Academy of Science's assessment of cumulative environmental effects of oil and gas activities on Alaska's North Slope (NAS 2003). The development of an oil and gas field in a wilderness area involves construction of roads, drilling pads and other infrastructure, waste disposal, electricity generation, and other activities each of which contributes its own effects to the cumulative effects. The cumulative effects analyses range from a reasonably simple addition of the areas of habitat lost due to all construction to estimation of reduced caribou calving due to the cumulative effects of all activities on behavior and stress.

29

Background and Reference

To determine how polluted the environment is or how much people or other organisms and communities are affected, you need an appropriate comparator. This is not a problem with experimental studies which include controls, but in the field, you will frequently have difficulty determining what comparisons are appropriate. Several years ago, I reviewed a study of effects of mine drainage on a lake, led by a respected consultant. He could not find uncontaminated sites with conditions equivalent to the contaminated areas and admitted that there was no real reference site. Nevertheless, he published an assessment which concluded that there were no effects based on comparison of contaminated and less contaminated sites that had different types of habitat. While I sympathized with his difficulty, I would not use his results.

29.1 Definitions

Background: A background site or condition has chemical concentrations or biological and physical conditions that are characteristic of the absence of human disturbance. That is a true natural background condition. Because natural conditions are rare in many regions, relaxed definitions may be used. Minimally affected background is conditions in the absence of significant human disturbance. Least disturbed background is the best available conditions given the present disturbance of the region, landscape, or habitat type.

Reference: A reference site or condition is equivalent to background in that there are limits on contamination or disturbance. However, while background is a state of nature that is independent of use in research or an assessment, reference sites of conditions are based on some purpose. Reference may be characterized in two ways. One is counterfactual. A site that is the same as the site being assessed, except for the contaminant or disturbance of concern, is a reference site. The reference need not be free of other contaminants or disturbances. That is, the subject site and reference site would be effectively the same but for the contaminant or disturbance of concern. The second definition is an aspirational reference. A site that represents the desired future state of a contaminated or disturbed site is a reference site. It could be a model site for a mitigation plan.

Positive reference: Positive references are sites that resemble the site being assessed including the agent being assessed or a hypothesized cause of an

DOI: 10.1201/9781003156307-32

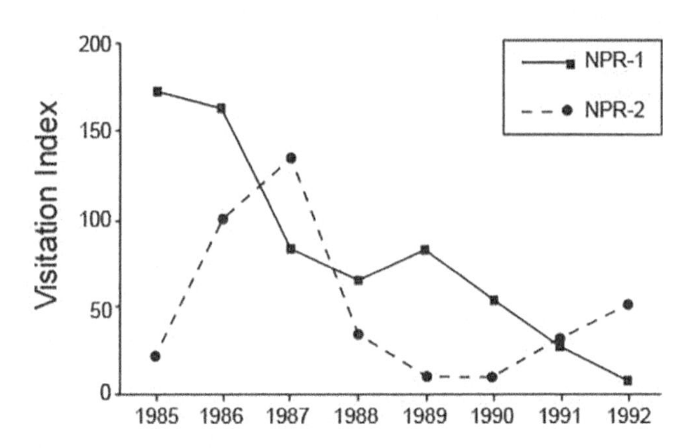

FIGURE 29.1
This figure illustrates the use of a positive reference site. It shows the decline of San Joaquin kit foxes on the NPR1 oil field, while foxes on the nearby and similarly disturbed and contaminated NPR2 oil field showed no consistent trend. This was evidence against climate and oil field development as causes of the NPR1 trend (Suter and O'Farrell 2009).

observed effect. Because the use of positive reference is not generally recognized, I will provide two examples. First, when assessing the Poplar Creek embayment in Oak Ridge, mercury was a contaminant of concern. We compared the embayment to uncontaminated embayments (negative references) but also to the more mercury-contaminated East Fork Poplar Creek (positive reference). If an effect on fish in the Poplar Creek embayment did not also occur in fish in the creek, that was evidence that mercury was not the cause. Second, in the San Joaquin kit fox study, the regular decline in foxes on Naval Petroleum Reserve 1 (NPR1) was not observed in the nearby and similarly developed NPR2 (Figure 29.1). NPR2 was a positive reference in that it shared NPR1's oil development and climate. The large differences in temporal dynamics of the fox population on the sites provided evidence against both climate and oil development as causes of the decline on NPR1. Something else was causing the difference between the two oil fields (Suter and O'Farrell 2009).

Properties of both background and reference sites may be defined for a location or a region. However, reference values are most often location specific, and background is most often region specific.

29.2 Uses

Background levels (i.e., concentrations at background sites) may be used as benchmark values for concentrations of naturally occurring chemicals in a political entity or region. That is, the goal may be to achieve or protect a natural uncontaminated and undisturbed condition. Screening values or regulatory limits for contaminants that are naturally occurring may be set at or near the upper bound of background concentrations for a region or a political entity.

Background or reference levels may be used as remedial goals. You may wish to restore pre-contamination levels in water, sediment, or soil, but you cannot expect to make those media cleaner than background. If a benchmark value is lower than a local background level, the benchmark may be adjusted upward to the background level.

Reference sites may be used as comparators to estimate the degree of contamination or effects at a contaminated site or water body. Comparison to background can exaggerate the influence of a source if regional conditions are contaminated or disturbed and the regional biotic community is degraded relative to background. In such cases, a local reference condition is the appropriate comparator. However, if a source pushes aqueous concentrations of a chemical over the ambient water quality standard, that is not acceptable in the U.S. To bring ambient concentrations below the standard, a total maximum daily load must be derived for each source of the chemical.

29.3 Identification of Background

There are many methods for identifying background sites or conditions, but none are generally accepted. We can say that background displays one or more of the characteristics of background, and we can identify background sites or conditions by weighing the evidence for those characteristics (Chapter 25) (Cormier et al. 2018).

- Background is characterized by low contaminant concentrations.
- Background is characterized by occurrence of taxa known to be sensitive to the contaminants or disturbances of concern.
- Background occurs in natural, undisturbed sites with no detectable sources of contamination.
- Background is characterized by inflections, break-points, or polymodality in distributions of concentrations.
- Background concentrations may be characterized by concentrations in natural source materials. Examples include using concentrations in headwater springs as background for surface waters or concentrations in surficial geological materials as background sources of sediments.
- High concentrations in a background area associated with natural anomalies such as hot springs, salt springs, or ore bodies can be eliminated from background data.

- Background may be characterized by differences in isotopic composition of natural and unnatural materials or differences in the "fingerprints" of chemical mixtures (the relative concentrations of constituents) from different sources. For example, polyaromatic hydrocarbon mixtures from natural oil seeps will differ from contaminants such as bunker fuel or diesel spills.
- Background for naturally occurring elements may be a concentration estimated by empirical hydro-biogeochemical modeling using data from minimally affected sites in a region (Olson and Hawkins 2012).

29.4 Identification of Reference

Negative reference sites are intended to characterize conditions in the absence of the source or contaminants of concern. For example, concentrations of a contaminant and biological metrics such as the number of species downstream of an effluent may be compared to an upstream reference. Alternatively, the reference may be determined by chemical or biological measurements on a gradient away from a source. Ideally, the reference would be the asymptote of an exposure or effects metric along that gradient. If there is no upstream–downstream relationship or gradient away from a terrestrial source, you must search for similar sites that are uncontaminated by the source of interest. Finally, some agencies have designated reference streams or other reference sites that may be used by assessors.

As described in the opening paragraph, none of these approaches to identifying a reference may apply well to your case. Differences between contaminated and reference sites other than the contaminate should be small relative to the magnitude of effects. For example, if a contaminated stream has no fish but all potential reference sites do, the effect is clear enough that the differences in flow, substrate, etc. can be neglected.

In general, it is desirable to have multiple reference sites. These can be replicates or different types of references. For example, in my study of the cause of San Joaquin kit fox decline discussed above, the contamination and most disturbance was located in the hills at the center of the NPR1 petroleum reserve. My local reference site was the less contaminated and less disturbed and flatter periphery of NPR1. Foxes there were much less exposed to oil development, but the habitat was ecologically distinct. The positive reference was the nearby and similar NPR2 oil field with similar habitat and where the fox population had varied but not consistently declined. A third type of reference was two areas with no fox declines and no oil production. Each type of reference provided a different type of comparison to provide a different type of evidence (Suter and O'Farrell 2009).

Reference values may be developed from characterization of a region rather than sites. For example, in regional surveys of random stream locations, the 25th centile of a concentration distribution from all sampled sites has been used as the upper bound for reference status (Herlihy and Sifneos 2008). Sites within the region could then be compared to that regional reference.

30

Assessors in the Courtroom

Your work as an assessor may lead to involvement in legal proceedings. I have testified before on an Administrative Law Court and have been deposed for two cases that were settled out of court. Some of my colleagues have spent much of their careers testifying on behalf of employers or clients—it can be a lucrative form of consultancy. Because out-of-court settlements are common, you are more likely to be deposed than to testify in court. However, depositions are legal proceedings—you will be sworn in, you will be questioned by attorneys for the other side, and attorneys for your side will object to inappropriate questions. Depositions also last longer than testimony in court because the attorneys questioning you are fishing for evidence.

In any legal proceeding, your attorneys should brief you in advance. Their most important advice is to tell the truth but no more truth than is needed to answer the question. It is tempting to elaborate or provide background information that you think would be helpful, but that can get you in trouble. It is even worse to speculate or guess. If you do not really know the answer, say so.

You may testify in either or both of two roles. You may be a technical expert witness or you may be a lay witness.

30.1 Technical Expert

If you are testifying as an expert witness with respect to assessment methods or a point of science, it is useful to understand legal standard for admissibility of technical testimony. In the U.S. federal courts and many state courts, the current standard is the Daubert rule, set by the Supreme Court of the United States in 1993. That rule declares judges to be gatekeeps of technical expert testimony. They determine whether the evidence is relevant and reliable and whether it qualifies as scientific knowledge based on the scientific method. The Daubert rule provides illustrative factors for judging the science:

1. whether the theory or technique in question can be and has been tested;
2. whether it has been subjected to peer review and publication;

DOI: 10.1201/9781003156307-33

3. its known or potential error rate;
4. the existence and maintenance of standards controlling its operation; and
5. whether it has attracted widespread acceptance within a relevant scientific community.

The rule assumes judges are capable of evaluating those standards. For number 1, test your methods when possible. Our conductivity benchmark was developed using West Virginia's data and tested with Kentucky data; the values were very close. Number 2 is easy, if your work has in fact been peer reviewed and published. In a case in which the method for deriving the conductivity benchmark was challenged, the mining company's consultants were presenting their unreviewed report, while the USEPA's benchmark was published in a report that was reviewed by the Science Advisory Board and in a set of five papers published in a peer-reviewed journal (USEPA 2011; Cormier and Suter 2013a; Cormier and Suter 2013b; Cormier, Suter et al. 2013a; Cormier, Suter et al. 2013b; Suter and Cormier 2013). The judge was impressed by the number of publications in a high impact journal and dismissed the testimony of the mining company's expert. So, it is always good to have peer-reviewed publications. If your publications are attacked by the other side (e.g., in a letter to the editor), write a response. That will help to maintain your reputation and prepare you for any legal challenge. See, for example, our rebuttals to fallacious attacks on the conductivity benchmark (Cormier and Suter 2013c; Cormier, Suter et al. 2020). Number 3, error rates are frequencies of errors per application of a method, which can be tested. This standard is aimed at forensic methods such as tire prints and lie detectors that have high error rates. Assessment methods cannot in general be tested against true risks. However, some estimation methods such as QSARs and read-across can be tested. Numbers 4 and 5 can be satisfied if you are following standard methods and frameworks. If the other side is considering a Daubert challenge to your testimony, they are likely to ask leading questions on these factors during your deposition.

30.2 Lay Witness

You may be called to testify about nontechnical aspects of how the assessment was generated or how decisions were made. For example, in the Bristol Bay, Pebble Mine case, the mining company's attorneys contended that the meetings of our assessment team were illegal because the mining company and the public were not invited to attend. I was questioned about the nature of the assessment team and its meetings. I was also questioned about my contacts during the assessment process, my membership in environmental groups, and my objectivity and biases. They questioned me about the contents of my emails, which they had obtained. (Never put anything in an email or on social media that you would not want to be interrogated about.) Only after all that did they question me about technical points in the assessment.

31

The Future

We are drowning in information, while starving for wisdom. The world henceforth will be run by synthesizers, people able to put together the right information at the right time, think critically about it, and make important choices wisely.

—*E.O. Wilson (1998)*

It is traditional to end a book of this sort with predictions of where the field is going. Since I lack the gift of prophesy, I will extrapolate current trends into the future. I can only hope that 50 years of watching trends in environmental science come and go gives me some basis for extrapolation. I will also throw in some wishful thinking—things that I hope will happen. There are three aspects of the field's future: environmental assessment problems to be solved, methods and tools for solving the problems, and the professionalization of environmental assessment.

31.1 Environmental Problems

Some environmental problems are far from solved and will require more and better assessments in the future.

31.1.1 Climate Change

Climate change is a bigger environmental problem than the others combined. I am, however, not an alarmist who believes that the earth will become uninhabitable. We humans are tough and our ability to create new technologies makes us adaptable to a degree that no other species can begin to match. Nonhuman species are not so fortunate.

Assessors must address two aspects of climate change. The first is changes in the climate that cause changes in habitats that result in effects on organisms and populations. The second is the environmental side effects of new and expanding technologies as humans attempt to limit climate change and adapt to the changes that occur.

The most prominent climate assessments are those of the Intergovernmental Program on Climate Change and, in the U.S., the National Climate Assessment.

DOI: 10.1201/9781003156307-34

These are mostly devoted to the proximate effects of planetary warming such as elevated temperatures, increased or decreased rain or snow, sea-level rise, loss of ice, and changes in the frequency and severity of hurricanes and other storms. These predictions are difficult enough, but they are relatively easy because they are largely physics. The climatic changes cause local weather effects such as heat waves, droughts, and floods. They in turn severely alter or destroy ecosystems and human communities and livelihoods.

Effects on human health are complicated by behavioral factors like installing air conditioners to avoid heat exhaustion or heat stroke, evacuating to avoid drowning, and moving to less hazardous regions. Health assessments should also consider more subtle effects on physical and mental health due to the effects of physiological and emotional stress.

Ecological assessments of climate change must consider how nonhuman organisms will survive without air conditioners, medical care, habitat maps, or automobiles to take them to new habitats. They may extend their ranges to the north or to higher elevations, they may find refugia that are less influenced by climate change, or they may acclimate or adapt to changed conditions. However, because of the interdependencies among species in ecosystems, those strategies may fail. If a species moves north, it will not succeed if the species that it consumes, that pollinate it, or are otherwise essential to it do not move with it or do not move quickly enough. Which species will successfully respond to climate change and which will need some human help to move to a new habitat? It is commonly assumed that generalists will get by but specialists probably will not (Pennisi 2022; Hanson 2021). That may be a reasonable generalization, but we really do not know what traits must be generalized for species to persist during climate change. These are important problems for resource management agencies, conservation organizations, and their assessment scientists (Moore and Schindler 2022).

Climate change creates problems for causal assessments. It is important to know what health and ecological effects are caused by climate change and which are caused by other agents or by interactions between climate and other agents like mosquitos. Climate change is causally difficult because it is ubiquitous, but it manifests differently in different regions and seasons. That means that even the most basic characteristic of causation, spatial-temporal association, is often difficult to demonstrate. This is another problem that calls for weighing of multiple types of evidence (Chapters 24 and 25). It will require some effort to determine what types and characteristics of evidence particularly implicate climate change as a cause.

The technologies that will replace fossil fuels have their own environmental hazards that must be assessed. Windmills kill birds and bats, concentrating solar power plants burn birds that fly into their beams, and even photovoltaic panels can kill birds (NWCC 2010; Ho 2016; Kosciuch et al. 2020). Inherently safe nuclear power would still generate radioactive wastes. Hydroelectric dams destroy riverine ecosystems and block fish migrations. New technologies are required for arctic and subarctic regions with extremely long nights and inadequate winds or small electrical grids. All zero-carbon technologies require metals that must be mined. Batteries and solar cells require rare elements. Currently, much attention is being directed to environmental effects from lithium but battery technologies keep changing and soon we may be worrying about vanadium for redox flow batteries.

31.1.2 Plastics

Plastics are an environmental problem with no easy solution. First, there is no mandate in the U.S. or most other countries to regulate the production, use, or release of plastics. Second, the diversity of plastic compounds, product forms (bottles, films, bags, toys, fishing gear, clothing, etc.), and waste forms make them an ill-defined problem. Organic polymers are inherently difficult to recycle. The literature on risks from plastics is incoherent and often anecdotal. In the U.S., the Save Our Seas 2.0 Act of 2020 deals with waste plastics, but still does not provide a regulatory mandate. It does, however, provide funds for water treatment and encouragement of better waste management, particularly for reuse and recycling. More pertinent to assessors, it provides funds for research and reviews of health and ecological effects of microplastics, microfibers, derelict fishing gear, and plastic waste in general. That information mandate may provide a basis for a new generation of environmental assessments of plastics.

Plastics contain chemicals added as plasticizers and stabilizers. They include phthalate esters and bisphenol A which have been implicated as endocrine disrupters. They leach from plastics and can be tested and their fate can be modeled like other soluble chemicals. When I began working on my first contaminated site in 1986, I was concerned about the phthalates that were appearing in all water samples from the site. They were leachates from plastic sample containers or the plastic tubing in the analytical instruments. No assessor would be so unknowing these days, but the effects of those chemicals are still unclear.

31.1.3 Endocrine Disrupters

Ever since Theo Colburn's book, *Our Stolen Future*, endocrine disrupting chemicals have been a public concern. Most chemicals that are described as endocrine disrupters bind to a receptor for an endocrine hormone and act as the hormone (agonists) or interfere with binding to the receptor (antagonists). Broad definitions of endocrine disruption include chemicals that promote or inhibit synthesis, secretion, transport, binding, action, or elimination of hormones. The USEPA has an active program that screens chemicals for estrogen, androgen, and thyroid activities. It has, however, been slow to roll out a testing program for endocrine disruption that would support risk assessment, despite legal mandates. The European Union has also been slow to regulate endocrine disrupters although some member countries are anxious to see action. At some point, environmental agencies must move forward with a program to specifically regulate endocrine disrupters or stick with the current approach. That is, they could just assume that if the endocrine system of humans and nonhuman animals is significantly disrupted, it would be caught by the standard tests and endpoints. Assessors will need to be involved in developing credible scientific support for any of these decisions.

31.1.4 Genetic Technologies

Genetic engineering is a scary idea for much of the public despite its ability to improve human health and food supplies and the absence of effects of genetically modified foods. Now a new form of genetic engineering called gene drives is likely to be even more frightening because they introduce traits that are spread through a population over multiple generations, even when it is a deleterious trait. The most discussed example is driving populations of *Anopheles* mosquitos to near extinction as a means of preventing malaria. It has also been proposed as a means to eliminate rats from islands where they are destroying native species. In other words, gene drives are potentially a species-specific substitute for chemical pesticides. That is very appealing. For example, most of the large trees on my land are ashes which are being killed by the introduced emerald ash borer. I would be glad to release ash borers with a gene drive that would eliminate the species from the Americas. However, there is a concern that borers bearing the gene drive would somehow make it back to Asia where it would reduce ash borers in their native habitat, and that somehow the lack of ash borers would damage the Asian environment. Some thought is being applied to environmental assessment of gene drives, but more work is required (Then, Kawall, and Valenzuela 2020).

31.2 Methods

Methods are needed to generate new data that are more useful for more problems. Methods are also needed for assessments that use the new data to better address the major environmental problems. Finally, tools for bringing data and assessment methods together more efficiently are needed to perform state-of-practice assessments in a timely manner.

31.2.1 Methods for Data Generation

New Approach Methods (NAMs) are being developed by the USEPA and other organizations as a response to two challenges. The first is the tens of thousands of chemicals in commerce, most of which have inadequate information concerning environmental toxicity and fate. The second is the cost and effort of conventional toxicology and the public pressures to eliminate whole-animal testing on ethical grounds. The USEPA and ECHA have plans for the development of NAMs (ECHA 2016; USEPA 2018b).

NAMs include:

- inference from chemical structures and properties—QSAR and read-across;
- cellular and molecular in vitro toxicity tests, also known as high-throughput testing;
- in vitro, high-throughput toxicokinetic studies;
- mechanistic mathematical models of toxicological processes;
- databases of conventional data for validation of NAMs.

The challenge now is to develop these tools to do more than screening and integrate them with traditional toxicology and chemistry data to perform risk assessments. For example, to use results of cell culture or molecular binding tests, we must extend exposure models from ambient concentrations to concentrations to which particular cell types are exposed.

New methods are needed to generate field data for environmental media, humans, and nonhuman organisms. Low-cost compact sensors are being developed and deployed but more are needed. Aerial and satellite monitoring platforms are also needed to monitor pollution sources and transport and responses of biotic communities. These methods, if deployed in the numbers required, will generate "big data" which will require techniques and facilities for receiving, organizing, storing, and providing access to the data.

31.2.2 Methods for Assessment

Having spent time since retirement reviewing the state of environmental assessment, I feel that we have a surfeit of assessment methods, although not necessarily the right ones. In addition to methods, we need information. For example, I cannot count the number of times that I have been asked how to estimate the sensitivity of a turtle species to some chemical. Some of this information may come from the NAMs discussed above, but we will never know enough to assess risks to all taxa from all agents in all conditions. We also need better access to and integration of methods and information. That may come from the tools and automation discussed below.

One area where we need methods is the human–nature interface. As discussed in Chapter 14, we need better methods to assess risks to the environment in a way that makes people care. In the U.S., we particularly need to make Congress and the White House Office of Information and Regulatory Affairs (OIRA) care. We have the value of statistical life for humans, but we do not have an equivalent accepted method for readily monetizing environmental endpoints. As a result, environmental effects often do not contribute to cost–benefit calculations.

31.2.3 Tools and Automation

The problems to be addressed by environmental assessment and particularly by ecological assessment are much too large for the available personnel and funds. I see no reason to expect that those limitations will change. An obvious response is more tools to facilitate assessment. Many are already available. For example, there are tools to facilitate systematic reviews of the literature and to develop QSARs and some other NAMs. However, assessment tools are scattered and have no common interface, no consistent terminology, metadata, or links to common assessment processes. As a result, a major need is to make assessment tools more available and easier to use.

My last activity with the Risk Assessment Forum (RAF) was to conduct a colloquium of the USEPA's ecological risk assessors. The goal was to define priority topics for the next round of RAF ecological projects. The participants had many ideas, but the clear first priority was to develop a system to make all of the Agency's ecological guidance, frameworks, methods, tools, expertise, and data available in a single coherent and useful form. The RAF Ecological Oversight Committee has been working on that since I retired, and I understand that they are nearly finished. Assuming that they are a success, their system could be a model for others and perhaps even an international system. Watch for it.

Tools and automations are even more needed for environmental impact assessment. Currently, in the U.S., the NEPA assessment process for projects requires an average of 4.5 years (CEQ 2018). That hurdle is holding up clean energy and conservation projects. Much of that delay is pure bureaucratic processes, but making the assessment process more efficient would help.

We also need tools for auditing research integrity to help ensure that our assessments are using good data. Journal editors now commonly use plagiarism detectors but other tools are needed. For example, image manipulation is a common cause of paper retractions, but it depends on expert examination of images. The currently available software for detection of image copying and manipulation, the most common reason for retraction, is unreliable (Cardenuto and Rocha 2022). Software to detect data set problems such as duplication, data with highly improbable variance (too good to be true), data that violate Benford's law (not enough small first digits), and impossible data (75% survival among 10 organisms) could be developed. I found many data errors when I was developing screening benchmarks because I was auditing the studies that I was using and because I was alerted to data issues by David Parkhurst of Indiana U. However, I still did not check all issues in the screening benchmarks data—I had not even heard of Benford's law. In cases where bad data get into the literature (Chapter 20), the coauthors, peer reviewers, and editors say that they do not have time or expertise to check the data.

Software to check data is being developed for some issues, but a validated general data checker would really help.

Tools for auditing broad areas of science or assessment are more ambitious than those for checking particular bad practices like plagiarism. For example, a toolkit for defining bad practices in epidemiology has been published, but it is just a long table of bad methods or practices and their implications (Soskolne et al. 2021). As the authors acknowledge, this toolkit would benefit from the creation of an application to make it useable. The same is true of sets of good and bad practices in other types of assessments and in this book.

I believe that all of these methods and tools could benefit from developments in artificial intelligence. Basic AI has been applied to literature searching, QSAR development, and a few other tasks. Much more powerful AI is becoming available, and we should consider how it can help. Imagine an AI tool that would take the genotype of any species and the structure of a chemical and determine the effects.

31.3 Professionalization of Environmental Assessment

As E.O. Wilson wrote, the world will belong to wise synthesizers. We assessors must be synthesizers and we should be wise. In addition, we should be proud to be assessors. We should take our profession seriously and practice it proudly because we are the ones who are in a position to make the environmental decisions beneficial, harmful, or even embarrassing. We must be synthesizers of the relevant sciences and practices of biology, chemistry, toxicology, engineering, philosophy, statistics, mathematics, social science, policy, law, and rhetoric. We should perform the syntheses with integrity and devotion to our tasks. Currently, most of the major decision-makers in the USEPA are lawyers, with a few environmental engineers, toxicologists, and chemists thrown in. If assessors perform their roles well, the profession of assessor will be recognized, and it will become apparent that assessors are well prepared for environmental decision making.

Every technology has its effects on the environment. Although environmental assessment has made tremendous progress during my lifetime, there are always new challenges. Even if there were no new pollutants or hazardous technologies, we can always do a better job of assessing the old ones. If some day you are a retired environmental assessor like me, you should be able to look back on a career that manifestly made the world better.

Part IV

Supporting Material

This is the stuff that is stuck at the end: the glossary, abbreviations and acronyms, and references. In addition, I have provided suggestions for student projects. When I was a student, I thought that the questions at the end of chapters were pretty lame and did not help with learning. I still think that and I believe that assessors learn best by doing. If you read the student projects suggestions, you will learn how a project for my introductory ecology course at U.C. Davis launched me into the real world of applying assessment to improve the local environment. I hope that a course based on this book will similarly inspire current students.

DOI: 10.1201/9781003156307-35

Student Projects

The principles in this text should prepare you to assess environmental conditions, their causes, risks, and outcomes. There are always environmental questions and concerns to which you could apply environmental assessment. You might even review or assess some national and international environmental issues or develop assessment tools. Such real-world needs could be the subjects of student projects. These projects might be tackled by one student, but I expect most of them will require teams.

When I was a doctoral student in ecology at University of California, Davis, in the early 1970s, the City of Davis was awakening to environmental concerns. I was a member of a group of graduate students who chose an assessment of environmental conditions and trends in Davis as our project in the Introduction to Ecology course. I chose land conversion and related agriculture issues as my topic. We were part of the group that organized a nascent green party called the Greater Davis Planning and Research Group. Our ecology course project was a condition assessment which served as a basis for that group's platform. In 1972, three new city council candidates were elected with the group's backing who went on to make Davis a green city. I was chosen to represent Davis on the Yolo County Water Resources Board.

A student assessment need not be so politically ambitious. In the early 1970s, green politics was new. However, environmental assessment can always make a difference. Currently in my community, Duke Energy is planning to install a major natural gas pipeline. It was strongly opposed by NOPE (Neighbors Opposed to Pipeline Extension), but they had no coherent assessment of the risks of fire or explosion and their potential consequences, ecological damage from construction, relative risks of different pipeline routes, comparative risks from railroad transport, etc. It would have been a good student project.

Another possibility is to critique an environmental assessment. In the U.S., federal assessments are subject to public comment (search the Federal Register). State agency assessments may also be available in your state. An assessment that becomes available for comment at the right time could benefit from objective and thoughtful comments and suggestions from a group of students. Ideally, suggestions would be supported by student analyses, not just comments that the agency should have done a particular analysis. The agency that derived the assessment should be willing to share their raw data.

Instead of critiquing a real assessment when it is open to comments, students may find one that is complete but of interest in your area or to the class and perform an outcome assessment. In such cases, you can determine the outcome including the decision made, its implementation, and the results. You can then consider how the assessment might have been more complete, informative, or predictive. Those findings could be used to determine how a different assessment might have changed the outcome for the better and how similar future assessments could be improved.

Another possibility would be developing ecological or human health benchmark values for chemicals that do not have a benchmark value or one that has not been updated recently. My group at ORNL developed ecological screening values for water, sediment, and soil in the 1990s. Those values are out of date, but they are still used because, for some chemicals, there are no others. On the human health side, the USEPA's provisional peer-reviewed toxicity values are a model for developing a benchmark on a short schedule (USEPA 2022c). New screening values that are up to date and rigorously derived would be a good project. If they are well done, they could be published in a journal.

If your school has researchers in environmental science who are developing new information in toxicology, environmental chemistry, or other relevant fields, students may collaborate with the researcher to demonstrate the potential application of their results to problem solving. At U.C. Davis, one of the ecology professors was Robert Rudd who had done pioneering work on the effects of pesticides on wildlife. In particular, he identified a new phenomenon, biomagnification, that accounted for the deaths on western grebes on Clear Lake, California, even though the aqueous concentrations of DDT were low. A group of us who were taking an ecological modeling course turned his data and other data into a mathematical simulation of the dynamics of DDT in the Clear Lake ecosystem.

The ongoing effects of climate change provide opportunities for environmental assessment. What environmental effects can you detect or predict? What are appropriate adaptive or protective strategies? Contact the relevant local or state agencies and ask how student projects might help.

Creative professors and students could come up with other ideas. The point is to make student projects relevant and useful. I believe that such projects are more instructive than answering lists of questions that I might provide.

Abbreviations and Acronyms

aka	also known as
BMD	Benchmark Dose
BN	Bayesian Network
CAA	United States Clean Air Act
CERCLA	Comprehensive Environmental Response, Compensation, and Liability Act (Superfund)
Ch	Chapter
DOE	United States Department of Energy
DPSIR	Driver-Pressure-State-Impact-Response
EC_{10}	Expected Concentration affecting 10% of organisms or species
HC_5	Hazardous Concentration for 5% of species in a species sensitivity distribution
ECHA	European Chemicals Agency
EIA	Environmental Impact Assessment
EIS	Environmental Impact Statement
EMAP	USEPA's Environmental Monitoring and Assessment Program
EFSA	European Food Safety Authority
ERA	Environmental or Ecological Risk Assessment
EU	European Union
FONSI	Finding of No Significant Impact
HC_x	Hazardous Concentration to x percent of species
HRS	Superfund Hazard Ranking System
IEAM	Integrated Environmental Assessment and Management
IPCC	Intergovernmental Program on Climate Change
IRIS	Integrated Risk Information System (USEPA chemical health risk assessments)
LC_{50}	Median lethal concentration
LD_{50}	Median lethal dose
MoE	Margin of Exposure
NAMs	New Approach Methods
NARS	USEPA's National Aquatic Resource Surveys
NAWQC	U.S. National Ambient Water Quality Criteria
NEPA	U.S. National Environmental Policy Act
NPL	Superfund National Priorities List
OECD	Organization for Economic Cooperation and Development
OIRA	White House Office of Information and Regulatory Affairs
ORD	USEPA's Office of Research and Development
ORNL	DOE's Oak Ridge National Laboratory
OW	USEPA's Office of Water
PAF	Potentially Affected Fraction

RAF USEPA's Risk Assessment Forum
REACH Registration, Evaluation, Authorization, and Restriction of Chemicals (European law)
Sec Section
SEM Structural Equation Model
SETAC Society for Environmental Toxicology and Chemistry
SSD Species Sensitivity Distribution
USEPA United States Environmental Protection Agency

Glossary

Some of these terms are not used in the text. However, they are likely to come up in the course of performing or reading environmental assessments or relevant information sources. The definitions provided here are the ones that are relevant to environmental assessment.

abduction: Inference to the best explanation. An alternative to *deduction* and *induction*.

accuracy: Closeness of a measured or computed value to the true value.

acid deposition: A process by which gaseous emissions of sulfur and nitrogen oxides are converted in the atmosphere into acid gases or particulates and either dry or wet deposited on the landscape.

acute: Occurring within a short period of time relative to the life span of an organism (conventionally <10%). Acute is also used to refer to severe effects, usually death.

adaptive management: The use of management actions as experimental treatments to test assessment models and thereby provide a better basis for future management actions. More generally, improving environmental management by learning from experience.

advocacy science: Scientific studies performed for the purpose of supporting a particular action, product, or sponsor.

agent: Any physical, chemical, or biological entity or process that can potentially cause an effect. It is synonymous with *stressor* but more general because it includes nutrients, water flow, and other agents that may be beneficial or neutral rather than stressful.

ambient media toxicity test: A toxicity test conducted with an environmental medium (soil, sediment, water). Usually, the media contain multiple contaminant chemicals.

analogy: An inference from similarity of known attributes to similarity in other attributes. In causal assessments, similar causes are expected to have similar effects and similar effects are expected to have similar causes.

analysis of exposure: A phase in an ecological risk assessment in which the spatial and temporal distribution of the intensity of the contact of endpoint entities with contaminants is estimated.

analysis of effects: A phase in an ecological risk assessment in which the relationship between exposure to contaminants and effects on endpoint entities is estimated. It may also include identification of the hazard.

analysis plan: A plan for performing an assessment, including the data to be collected and the modeling and other analyses to be performed to provide the needed input to the environmental management decision.

antagonism: The process by which two or more chemicals cause joint effects that are less than additive (either exposure additive or response additive).

antecedence: The characteristic of a causal relationship that connects a proximate cause to processes that precede it.

assessment endpoint: An explicit expression of the environmental value to be protected. An assessment endpoint must include an entity and an attribute of that entity.

assessment exposure: The form and measure of exposure that is estimated by an assessment. It is the form of exposure that will be the subject of regulation or remediation. See *measurement exposure.*

assessment factor: Factors by which a value is divided to account for extrapolations and uncertainties.

assessor: An individual engaged in the performance of environmental assessments.

association: The degree to which one variable (e.g., representing a cause) co-occurs or covaries with another (e.g., representing an effect). Correlation is a measure of association.

background concentration: The concentration of a substance in an environmental medium that is not contaminated by the sources being assessed or any other sources. Background concentrations are due to natural occurrence.

baseline assessment: A risk assessment that determines the risks associated with current conditions so as to determine whether remediation is required.

Bayesian: A branch of statistics characterized by the updating of prior beliefs, estimation of conditional probabilities using Bayes theorem, and defining probabilities as subjective degrees of belief.

benchmark: A generic term for values that delimit levels of an agent that are in some sense acceptable and those that suggest the need for testing, assessment, regulation, or other action. The term encompasses standards, criteria, guidelines, objectives, screening values, concern levels, and intervention levels.

benchmark dose: A dose of a substance associated with a specified low level of an effect (usually 10%) divided by assessment factors. The term is used in the U.S. EPA's human health risk assessments and sometimes risk assessments for avian and mammalian wildlife.

bias: A systematic deviation of measured or computed values from true values. Alternatively, a person's propensity to favor a particular outcome.

bioassessment (biological assessment): Evaluation of ecosystem condition using biological surveys and other direct measurements of resident biota.

bioassay: A procedure in which measures of biological responses are used to estimate the concentration or to determine the presence of some chemical or material. See *toxicity test*.

bioaccumulation: The net accumulation of a substance by an organism due to uptake from all environmental media.

bioaccumulation factor: The quotient of the concentration of a bioaccumulated chemical in an organism divided by the concentration in an environmental medium. The concentrations should be near steady state.

bioavailability: The extent to which a form of a chemical is susceptible to being taken up by an organism. A chemical is said to be bioavailable if it is in a form that is readily taken up (e.g., dissolved) rather than a less available form (e.g., sorbed to solids or to dissolved organic matter).

bioconcentration: The net accumulation of a substance by an organism due to uptake directly from aqueous solution.

bioconcentration factor: The quotient of the bioconcentrated concentration of a chemical in an organism divided by the concentration in water. The concentrations should be near steady state.

bioindicator: A species or group of species that, by their presence or abundance, are indicative of a property of the ecosystem in which they are found. Enchytraeid worms are bioindicators of low dissolved oxygen.

biomagnification: The increase in concentration of a chemical in a consumer species relative to concentration in food species in a food web.

biomagnification factor: The ratio of the biomagnified concentration of a chemical in organisms at a particular trophic level to the concentration at the next lower level.

biomarker: A measurable change in a biochemical, cellular, or physiological characteristic of an organism that may be used as an indicator or measure of exposure or effect.

biosurvey: A process of counting or measuring some property of biological populations or communities in the field. An abbreviation of biological survey.

biota-sediment accumulation factor: The ratio of the concentration of a chemical in a benthic organism to the concentration in sediment.

chronic: Occurring after a long period of time relative to the life span of an organism or effectively infinite in duration relative to the response rate of the exposed system. Chronic is also used to refer to nonlethal effects or effects on early life stages, but that usage causes confusion.

coherence: The quality of a body of evidence that its constituent pieces are logically linked, thus forming a reasonable explanation.

co-occurrence: (1). The characteristic of a causal relationship that the cause and effect are collocated in space and time. (2) An instance of collocation in space and time.

combined effect: An effect of a mixture of agents.

community: A biotic community consists of all plants, animals, and microbes occupying the same area at the same time. However, the term is also commonly used to refer to a subset of the community such as the fish community or the benthic macroinvertebrate community. The latter may more properly be termed an assemblage.

comparative risk assessment: Risk assessments used to rank or otherwise compare alternative actions to address a particular risk, to prioritize risks for remedial or regulatory action, or to compare chemicals or other agents for a particular use.

compensation: In population ecology, compensation is the increase in growth of a population at low densities due to decreased mortality, more rapid growth and maturation, and increased fecundity. In ecosystem ecology, compensation is the increased rate of performance of a process by one or more species as the abundance or activity of other species declines. For example, increased growth and mast production by chestnut oak compensated for the loss of American chestnut trees in southern Appalachian forests.

concentration additivity: A mode of combined toxicity in which each chemical behaves as a concentration or dilution of the other, based on their relative toxicities.

conceptual model: A representation of the hypothesized causal relationships between the sources of a contaminant or other agent and the responses of the endpoint entities. It typically includes a diagram and explanatory text.

concordant: A property of a model having consistent units, and for probabilities or frequencies, having units with respect to a common base (e.g., per year).

confounder: Agents or influencing factors that interfere with the ability to quantify the contribution of a specific cause to an observed biological effect.

confounding: Bias in the statistical representation of a causal relationship due to the presence of a confounder.

consideration: A property, type, or implication of evidence used to evaluate the weight of evidence.

contaminant: A substance that is present in the environment due to release from an anthropogenic source. If it is believed to be potentially harmful, it may be termed a *pollutant*.

corroboration: Supporting evidence for a candidate cause from one or more independent studies providing similar results.

cost–benefit analysis: Methods for balancing the costs and benefits associated with an action or technology; aka *benefit–cost analysis*.

cumulative effect: The combination of multiple effects.

deduction: Inference from a general theorem or set of axioms to a particular conclusion. Deductive arguments are valid if the conclusions are always true when the premises are true. (See *abduction* and *induction*.)

definitive assessment: An assessment intended to estimate the risks and provide the basis for management decisions. (See *screening assessments*.)

de manifestis: Sufficiently large to be obviously significant (i.e., risks so severe that actions are nearly always taken to prevent or remediate them).

de minimis: Sufficiently small to be ignored (i.e., risks low enough to not require actions to prevent or remediate them).

depensation: Depensation is the accelerated decline in a population at low densities due to reduced ability to find mates, increased predation, or decreased ability to condition the environment. It is the opposite of *compensation*.

detection limit: The concentration of a chemical in a medium that can be reliably detected by an analytical method. It is defined statistically (e.g., as the concentration that has a prescribed probability of being greater than zero, given variability in the analytical method).

deterministic: Having only one possible outcome.

direct effect: An effect resulting from an agent acting on an environmental component of interest, not through effects on other components of the ecosystem. See also *indirect effect* and *secondary effect*.

dose: The amount of a chemical, chemical mixture, pathogen or radiation delivered to an organism. For example, mg of Cd per kg of mallard duck (mg/kg) administered by oral gavage.

dose rate: The dose per unit time (e.g., mg/kg-d).

dose additivity: A mode of combined toxicity in which each chemical behaves as a concentration or dilution of the other based on their relative toxicities.

dredge spoil: Sediments dredged from a water body and deposited as waste at another aquatic location or on land.

ecoepidemiology: The analysis of the causes and consequences of observed effects on ecological entities in the environment.

ecological entity: An ecosystem, functional group, community, population, or type of organism which may be exposed to a hazardous agent or may itself be a hazardous agent.

ecological risk assessment: A process that evaluates the adverse ecological effects resulting from exposure to one or more agents.

ecosystem: The functional system consisting of the biotic community and abiotic environment occupying a specified location.

effect: (1) In general, an effect is some change in an entity that follows a cause. A biological effect is the biological result of exposure to a causal agent or event. This term is similar to response, but emphasizes the agent that acts (e.g., the effect of cadmium) rather than the receptor that responds to it (e.g., the response of trout). (2) In practice, an effect is an observed or predicted discrepancy of an entity from its expected or nominal condition (e.g., the number of species in a biotic community relative to reference communities).

emission: A discharge to the environment.

empirical model: A mathematical model that is derived by fitting a function to data using statistical techniques or judgment. Purely empirical models summarize relationships in data sets and have no mechanistic interpretation.

endpoint attribute: An attribute of an endpoint entity that has been chosen for protection and is estimated by an assessment. The endpoint attribute is one component of the definition of an assessment endpoint.

endpoint entity: An organism, population, species, community, or ecosystem that has been chosen for protection. The endpoint entity is one component of the definition of an assessment endpoint.

environmental assessment: (1) A process of generating and presenting scientific information to inform an environmental management decision. (2) The product of an environmental assessment process.

environmental impact assessment: The assessment of the environmental effects of a plan, policy, program, or actual projects to inform the decision to proceed or not. The term refers to either the process or an individual assessment. In the U.S., more complete impact assessments are referred to as environmental impact statements.

environmental risk: A risk to humans or other entities due to hazardous agents in the environment. This definition applies to the U.S., U.K., and some other nations. However, some nations use environmental risk equivalently to ecological risk.

epidemiology: The analysis of the causes and consequences of observed effects on human populations.

equilibrium partitioning: The transfer of a chemical among environmental media so that the relative concentrations of any two media are constant.

evidence: Information that can contribute to judging the truth of a hypothesis.

excess risk: The difference between the risk given an exposure and the risk without the exposure or with an alternative exposure.

exotic species: A biological species that has been introduced from elsewhere, including species produced by biological engineering, selective breeding, or natural selection.

exposure : The contact or co-occurrence of a contaminant or other agent with a biological receptor.

exposure pathway: The physical route by which a chemical or other agent moves from a source to a biological receptor. A pathway may involve transformations and exchanges among multiple media.

exposure profile: The product of characterization of exposure in the analysis phase of ecological risk assessment. The exposure profile summarizes the magnitude and spatial and temporal patterns of exposure for the scenarios described in the conceptual model.

exposure–response profile: The product of the characterization of ecological effects in the analysis phase of ecological risk assessment. The exposure–response profile summarizes the data on the effects of a contaminant, the relationship of the measures of effect to the assessment endpoint, and the relationship of the estimates of effects on the assessment endpoint to the measures of exposure.

exposure–response relationship: A quantitative relationship between the measures of exposure to an agent and a measure of effect. Exposure–response relationships may take various forms including thresholds (e.g., effects occur at concentrations greater than x mg/L), statistical models (e.g., the proportion dead as a probit function of concentration), or mathematical process models (e.g., dissolved oxygen concentration as a function of phosphorous loading and ecosystem variables). Dose–response, concentration–response, and time-to-death models are specific examples of exposure–response relationships.

exposure route: The means by which a contaminant enters an organism (e.g., inhalation, stomatal uptake, ingestion).

exposure scenario: A set of assumptions concerning how an exposure may take place, including assumptions about the setting of the exposure, characteristics of the agent, activities that may lead to exposure, conditions modifying exposure, and temporal pattern of exposure.

extirpation: Effective elimination of a species from an ecosystem, watershed, or region. A synonym is functional extinction.

extrapolation: (1) The use of related data to estimate an unobserved or unmeasured value. Examples include use of data for fathead minnows to estimate effects on yellow perch or use of data on oxidation rates in 20°C water to estimate rates at 5°C. (2) Estimation of the value of an empirical function at a point outside the range of data used to derive the function.

feasibility study: The component of the CERCLA (Superfund) remedial investigation/feasibility study that is conducted to analyze the benefits, costs, and risks associated with remedial alternatives.

frequency: The number of observations of an event or condition in a class. For example, the frequency of releases of untreated effluent from a treatment plant is two per year.

frequentist: A branch of statistics characterized by the analysis of a data set as one of a potentially infinite number of samples drawn from population with a particular distribution and by the treatment of probabilities as frequencies.

geophagous: Deliberately eating soil.

habitat: An area that provides the needs of a particular species or set of species.

hazard: A situation which may lead to harm. In risk assessment, a hazard is a hypothesized association between an agent capable of causing a particular effect and a potentially susceptible endpoint entity. Identification of a hazard leads to assessment of the risk that the harm will occur or the level of exposure at which the harm will not occur.

hazard quotient: The quotient of the ratio of the estimated level of an agent divided by a level which is estimated to have no effect or to cause a prescribed effect. For example, the concentration of a chemical in water divided by its LC_{50}. (Also, assessment quotient or risk quotient.)

human health risk assessment: A process that evaluates the adverse effects on humans resulting from exposure to one or more agents.

impairment: A detrimental effect on a medium, population, community, or ecosystem that is sufficient to prompt a management or regulatory action.

implication: The meaning of evidence with respect to a hypothesis. Information becomes evidence when it has implications for a hypothesis.

indicator: An indicator for ecological or health monitoring would be an observation that indicates something about the ecosystem or human population that is important, but not easily observed.

indirect effect: An effect resulting from the action of an agent on components of the ecosystem, which in turn affects the assessment endpoint or other ecological component of interest (see *direct effect*). Indirect effects of chemical contaminants include reduced abundance due to toxic effects on food species or on plants that provide habitat structure. Equivalent to *secondary effects*, but also includes tertiary and quaternary effects.

induction: In logic, induction is the derivation of general principles from observations. For example, a series of observations of bioconcentration of different organic chemicals may allow us to induce that the bioconcentration factor (BCF) is a function of octanol–water partitioning coefficients (K_{ow}); in particular, $BCF = 0.89 \log K_{ow} + 0.61$. Inductive arguments are valid if the conclusions are usually true when the premises are true. (See *abduction* and *deduction*.)

inference: The act of reasoning from evidence.

interaction: The characteristic of causal relationships that a causal agent contacts, impinges upon or enters a susceptible entity in a way that initiates the effect.

junk science: Scientific results that are said to be false because of perceived political, financial, or other motives other than a desire for truth. The term is itself political, having been developed by industry groups to discredit environmental and public health concerns. The antonym is *sound science*. A less politically loaded term is biased science.

kinetic: Referring to movement. In particular, in toxicology and pharmacology, kinetic refers to the movement and transformation of a chemical in an organism (i.e., *toxicokinetic* or pharmacokinetic).

life-cycle assessment: A method for determining the relative environmental impacts of alternative products and technologies based on the consequences of their life cycle, from extraction of raw materials to disposal of the product following use.

likelihood: The hypothetical probability that events had a prescribed outcome. It may be thought of as the probability of evidence given a hypothesis $[P(E|H_x)]$ or as the probability of a sample $(x_1, x_2, \ldots, x_n)$ given a probability density function. Likelihoods are termed hypothetical probabilities, because the sum of likelihoods across a set of alternative hypotheses may be greater than one. In ordinary English usage, it is the chance or expectation of occurrence.

loading: The rate of input of a pollutant or other agent to a particular receiving system (e.g., nitrogen loading to the Chesapeake Bay).

lowest observed adverse effect level (LOAEL): The lowest level of exposure to a chemical in a test that causes statistically significant differences from the controls in any measured response. LOAELs are no longer considered a good practice, but they are still in the literature.

margin of exposure: the benchmark effects level relative to the magnitude of the exposure level is the margin of exposure (MOE). A large margin of exposure indicates that exposure must be much higher to reach the benchmark.

measure of effect: A measurable or estimable human or ecological characteristic that is related to the valued characteristic chosen as the assessment endpoint (similar to but broader than the earlier term *measurement endpoint*). For example, a biochemical or behavioral change in an organism is a measure of effect but not an endpoint.

measurement endpoint: The value derived by a toxicity test or an equivalent study to express the effect of the agent on the receptor. Examples include LC_{50}, EC_{10}, or formerly used endpoints like a *no observed adverse effect level.*

measurement exposure: A measurable or estimable characteristic of a contaminant or other agent that is used to quantify exposure with respect to an exposure–response relationship. See *assessment exposure.*

mechanism of action: The specific process by which an effect is induced. It is often used interchangeably with *mode of action* but is usually used to describe events at a lower level of organization.

mechanistic model: A mathematical model that estimates properties of a system by simulating its component processes rather than using empirical relationships.

media toxicity test: A toxicity test of water, soil, sediment, air, or biotic media that is intended to determine the toxic effects of exposure to that medium. It includes *ambient media toxicity tests* plus tests of media that have been spiked or otherwise treated.

median lethal concentration (LC_{50}): A statistically or graphically estimated concentration that is expected to be lethal to 50% of a group of organisms under specified conditions.

mesocosm: A relatively large outdoor or indoor facility with controlled physicochemical conditions and multiple species used to physically simulate natural ecosystems.

meta-analysis: A set of statistical methods for combining the numerical results of multiple studies of the same problem to obtain a common judgment of a hypothesis or numerical estimate.

mode of action: A phenomenological description of how an effect is induced (see *mechanism of action*). For example, if the effect of interest is local extinction of a species, the mode of action might be habitat loss and the mechanism of action might be agricultural tillage.

model: A mathematical, physical, or conceptual representation of a system.

model uncertainty: The component of uncertainty concerning an estimated value that is due to misspecification of a model used for the estimation. It may be due to the choice of the form of the model, its component parameters, or its bounds.

Monte Carlo simulation: A resampling technique frequently used in "uncertainty" (actually, confidence) analysis in assessments to estimate the distribution of a model's output parameter from distributions of input parameters.

natural attenuation: Degradation or dilution of chemical contaminants by unenhanced biological and physicochemical processes.

net environmental benefits: The gains in environmental services or other ecological properties attained by remediation or ecological restoration, minus the environmental injuries caused by those actions.

no observed adverse effect level (NOAEL): The highest level of exposure to a chemical in a test that does not cause statistically significant differences from the controls in any measured response. NOAELs are no longer considered a good practice, but they are still in the literature.

normalization: Alteration of a chemical concentration or other property to reduce variance due to some characteristic of an organism or its environment (e.g., division of the body burden of an organic chemical by the organism's lipid content to generate a lipid-normalized concentration).

octanol–water partition coefficient (Kow): The quotient of the concentration of an organic chemical dissolved in octanol divided by the concentration dissolved in water if the chemical is in equilibrium between the two solvents.

phytoremediation: Remediation of contaminated soil via the accumulation of the chemicals by plants or the promotion of degradation by plants.

phytotoxicity: Toxicity to plants.

piece of evidence: The basic unit of evidence; examples include the results of a toxicity test, a quantitative structure–activity model, or a stream survey.

policy: A principle of action adopted or proposed by a government or other entity. For example, the USEPA has had a policy that mutagens are considered potential human carcinogens, unless refuted by evidence.

pollutant: A substance that is present in the environment due to release from an anthropogenic source and is believed to be harmful. See *contaminant.*

population: The aggregate of interbreeding individuals of a species occupying a location.

precision: The exactitude with which a measurement or estimate can be specified and reproduced, usually determined by the similarity of independent determinations.

preliminary remedial goal (PRG): A concentration of a contaminant in a medium that serves as a default estimate of a remedial goal for that medium.

primary data: Original unsummarized data.

probability: A scale from 0 to 1 that may be interpreted as the relative frequency of occurrence of an event in repeated trials or time intervals or as the degree of belief assigned to a hypothesis. The probability scale is interpreted as 0 indicating impossibility (subjectivist) or never occurs (frequentist) and 1 indicating inevitability (subjectivist) or always occurs (frequentist).

problem formulation: The phase in an ecological risk assessment in which the goals and environment of the assessment are used to define the assessment endpoints and the methods for achieving those goals.

pseudoreplication: The treatment of multiple samples from a single treated location or system as if they were samples from multiple independently treated locations or systems. For example, multiple samples of benthic invertebrates from a stream reach below a wastewater outfall are pseudoreplicates.

quantitation limit: The concentration of a chemical in a medium that can be reliably quantified by an analytical method. Statistical definitions differ, but are generally based on concentrations that can be estimated with prescribed precision (e.g., the true concentration that produces estimates having a relative standard deviation of 10%).

receptor: An organism, population, or community that is exposed to contaminants. Receptors may or may not be assessment endpoint entities.

record of decision: The document presenting the final decision resulting from the CERCLA remedial investigation/feasibility study process regarding selected alternative action(s).

recovery: The return of a population, community, or ecosystem to a state with the valued properties of a previous reference state.

reference, positive: A site or the information obtained from that site used to estimate the state of a system exposed to a contaminant other than the system that is being assessed.

reference, negative: A site or the information obtained from that site used to estimate the state of a receiving system in the absence of contamination. Reference, used without qualification, refers to a negative reference.

regulation: A regulation is a rule that specifies how a law will be implemented.

relative risk: The ratio of the risk given an exposure to the risk without the exposure or with an alternative exposure.

remedial alternative: One of a number of potentially applicable remedial technologies or actions proposed for a contaminated site.

remedial goal: A contaminant concentration, toxic response level, or other criterion that is selected by the risk manager to define the condition to be achieved by remedial actions.

remedial goal option: One or more contaminant concentrations, toxic responses, or other criterion that is recommended by the risk assessors as likely to achieve conditions protective of the assessment endpoints.

remediation: Actions taken to reduce risks from contaminants including removal or treatment of contaminants and restrictions on land use. Note that, in contrast to *restoration*, remediation focuses strictly on reducing risks from contaminants and may actually reduce environmental quality.

removal action: An interim remedy for an immediate threat from contaminants.

restoration: Actions taken to make the environment whole, including restoring the capability of natural resources to provide services to humans. Restoration goes beyond *remediation* to include restocking, habitat rehabilitation, reduced harvesting during a recovery period, etc.

riparian: Occurring in or by the edge of a stream or in its floodplain.

risk: The potential for adverse consequences for humans or ecological systems.

risk characterization: A phase of an environmental risk assessment that integrates the exposure and the exposure–response relationship to estimate adverse effects associated with exposure to the contaminants. It also addresses confidence in the estimates and uncertainties.

risk manager: An individual with the authority to decide what actions will be taken in response to a risk. Examples of risk managers include representatives of regulatory agencies, land managers, and safety officers.

risk management: The processes of deciding whether to accept a risk or to take actions to reduce it and of justifying and implementing the decision.

scenario: A possible future condition, given certain assumed actions and environmental conditions. In risk assessment, a scenario is a set of hypothetical or actual conditions under which exposure may occur and for which risks will be characterized.

scoping assessment: A qualitative assessment that determines whether a hazard exists that is appropriate for a risk assessment. For contaminated sites, it determines whether contaminants are present and whether there are potential exposure pathways and receptors.

screening assessment: A simple quantitative assessment performed to guide the planning of a subsequent assessment by eliminating agents, receptors, or areas from further consideration. They are intended to screen out certain issues rather than to guide a management decision. (See *scoping assessment* and *definitive assessment.*)

screening benchmark: A concentration or dose which is considered a threshold for concern in the screening of contaminants from further consideration in a risk assessment.

secondary data: Data obtained from the literature. Secondary data are not designed to meet the assessor's quality requirements or to estimate a particular assessment parameter or function, and it often includes only summary data.

secondary effect: An effect of an agent caused by effects on an entity that influences the endpoint entity rather than by direct effects on the endpoint entity. For example, herbicides kill plants (a primary or direct effect) which may cause loss of habitat structure and food resulting in reduced herbivore abundance (the secondary effect). See also *indirect effect, direct effect,* and *primary effect.*

sensitivity: (1) In modeling, the degree to which model outputs are changed by changes in selected input parameters. (2) In biology, the degree to which an organism or other entity responds to a specified change in exposure to an agent.

sentinel species: A species that displays a particularly sensitive response to a chemical or other agent or is highly exposed to a contaminated environment. This property makes them useful indicators of the presence of hazardous levels of the agent to which they are sensitive.

single chemical toxicity test: A toxicity test of an individual chemical administered to an organism or added to soil, sediment, air, or water to which an organism is exposed.

site: An area of land or water which is the setting for an assessment because (1) it has been identified as contaminated or disturbed and potentially in need of remediation or restoration, or (2) it is a planned location for a potentially hazardous activity.

source: An entity or action that releases contaminants or other agents into the environment (primary source) or a contaminated medium that releases the contaminants into other media (secondary source). Examples of primary sources for contaminated sites include spills, leaking tanks, dumps, and waste lagoons. An example of a secondary source is contaminated sediments that release contaminants by diffusion, bioaccumulation, and exchange. The term source is also used more generally to indicate the activities or drivers that are the sources of development, physical disturbance, or use.

species sensitivity distribution (SSD): A distribution function of the toxicity of a chemical or mixture to a set of species that may represent a taxonomic group, assemblage, or community.

stakeholder: An individual or organization that has an interest in the outcome of a regulatory or remedial action but is not an officially party to the decision making. Examples include natural resource agencies, local governments, and citizens groups. The synonym *interested party* is clearer but less commonly used.

stochasticity: Apparently random changes in a state or process that are attributed to inherent randomness of the system.

stressor: Stressor is commonly used in the U.S. in place of *agent*. It implies a prejudgment that the agent being assessed will have adverse effects. Just as the dose makes the poison, the level of exposure, the receptor and the environmental conditions make an agent a stressor.

sound science: Scientific results that are said to be credible. The term is usually used in a political context to describe results that support the speaker's positions. The political antonym is *junk science*.

strength: The degree to which evidence demonstrates a large difference or a high degree of association between a cause and effect relative to background levels. It is a property of evidence.

sufficiency: The characteristic of a causal relationship that the agent or event must be adequate to induce the effect in susceptible entities.

superfund: The common name for the Comprehensive Environmental Response and Liability Act (CERCLA). It is the law in the U.S. that mandates the assessment and, as appropriate, remediation of contaminated sites. The name comes from a fund that was created by taxing the chemical industry.

synergism: The process by which two or more chemicals or other agents cause joint effects that are more than additive (either exposure additive or response additive).

systematic review: A literature review that uses a predefined procedure to search the literature, screen the results, extract information, and document the process.

tertiary data: Data obtained from a published literature review or an electronic database of information derived from the literature. Like secondary data, tertiary data are not designed to meet the assessor's quality requirements or to estimate a particular assessment parameter. In addition, tertiary data may contain errors due to transcription or data entry and may not contain supporting information that is critical to interpretation.

time order: The characteristic of a causal relationship that the cause precedes the effect.

toxicity identification and evaluation (TIE): A process whereby the toxic components of mixtures (usually aqueous effluents) are identified by removing components of the mixture and testing the residue, fractionating the mixture and testing the fractions, or adding components of the mixture to background medium and testing the artificially contaminated medium.

toxicity test: A procedure in which organisms or communities are exposed to defined levels of a chemical or material to determine the nature and magnitude of responses. See *bioassay*.

toxicodynamics: The study of the processes by which exposure to a chemical or mixture induces a toxic effect or a description of the results of such studies. In particular, toxicodynamics focuses on the biochemical processes by which an internal exposure induces injuries.

toxicokinetics: The study of the processes by which an external exposure to a potentially toxic chemical or mixture (e.g., a concentration in an ambient medium or a dose) results in an internal exposure (e.g., concentration at a site of action), or a description of results of such studies.

type of evidence: A category of evidence used to characterize risk or identify a cause. The most common general types of evidence in ecological risk assessments of contaminated sites are (1) biological surveys, (2) toxicity tests of contaminated media, and (3) toxicity tests of individual chemicals, each of which is compared to chemical analyses of the media.

uncertainty: Lack of knowledge concerning an event, state, model, or parameter. Uncertainty may be reduced by research or observation.

uncertainty factor: A factor applied to an exposure or effect estimate to assure protection despite uncertainty.

uptake: Movement of a chemical from the environment into an organism as a result of any process.

uptake factor: The quotient of the concentration of element or compound in an organism divided by the concentration in an environmental medium. It is used interchangeably with *bioconcentration factor* and *bioaccumulation factor* but is most often applied to uptake from food or ingested water by terrestrial species.

variability: Differences among entities or states of an entity attributable to heterogeneity. Variability is an inherent property of nature and may not be reduced by measurement. Examples include the differences in the weights of adult fathead minnows or differences among years in the minimum flow of a stream.

water effect ratio: A factor by which a water quality benchmark is multiplied to adjust for site-specific water chemistry.

watershed: An area of land from which water drains to a common surface water body.

weigh: Consider the logical implications and properties of a body of evidence to assess the likelihood of a hypothesis.

weight: (1) (noun) The importance of a piece or category of evidence. (2) (verb) Assign importance to a piece or category of evidence.

weight of evidence: (1) A process of identifying the best-supported hypothesis (e.g., a cause or risk characterization) given the existence of multiple pieces of evidence or the results of such a process. (2) The relative degree of support for a candidate cause or other conclusion provided by evidence.

wildlife: Non-domestic terrestrial or semiaquatic vertebrates. Wildlife includes mammals, birds, reptiles, and amphibians.

References

Abrams, B., A. Anderson, C. Blackmore, F. J. Bove, S. K. Condon, C.R. Eheman, J. Fagliano, L.B. Haynes, L. S. Lewis, J. Major, M.A. McGeehin, E. Simms, K. Sircar, J. Soler, M. Stanbury, S. M. Watkins, and D. Wartenberg. 2013. "Investigating suspected cancer clusters and responding to community concerns, guidelines from CDC and the Council of State and Territorial Epidemiologists." *Morbidity Mortality Rep* 62 (8):1–28.

Adams, D. F. 1963. "Recognition of the effects of fluorides on vegetation." *J Air Pollut Control Assoc* 13:360–362.

Agerstrand, M., and A. Beronium. 2016. "Weight of evidence evaluation and systematic review in EU chemical risk assessment: Foundation is laid but guidance is needed." *Environ Int* 92–93:590–596.

Aldenberg, T., and J. A. Jaworska. 2000. "Uncertainty of the hazardous concentration and fraction affected for normal species sensitivity distributions." *Ecotoxicol Environ Saf* 46:1–18.

Ankley, G. T., R. S. Bennett, R. J. Erickson, D. J. Hoff, M. W. Hornung, R. D. Johnson, D. R. Mount, J. W. Nichols, C. L. Russom, P. K. Schmieder, J. A. Serrrano, J. E. Tietge, and D. L. Villeneuve. 2010. "Adverse outcome pathways: a conceptual framework to support ecotoxicology research and risk assessment." *Environ Toxicol Chem* 29 (3):730–41

ANZG. 2018. "Australian and New Zealand Guidelines for Fresh and Marine Water Quality." Australian and New Zealand Governments and Australian state and territory governments. www.waterquality.gov.au/anz-guidelines.

Arnold, S. F., D. M. Klotz, B. M. Collins, P. M. Vonier, L. J. Guillette Jr, and J. A. McLachlan. 1996. "Synergistic activation of estrogen receptor with combinations of environmental chemicals." *Science* 272 (5267):1489–1492.

Arnot, J. A., L. Toose, J. M. Armitage, M. Embry, A. Sangion, and L. Hughes. 2022. "A weight of evidence approach for bioaccumulation assessment." *Integr Environ Assess Manage* 18 (n/a) https://doi.org/10.1002/ieam.4583.

Arts, Gertie H.P., Laura L. Buijse-Bogdan, J. Dick M. Belgers, Caroline H. van Rhenen-Kersten, Rene PA van Wijngaarden, Ivo Roessink, Steve J Maund, Paul J van den Brink, and Theo CM Brock. 2006. "Ecological impact in ditch mesocosms of simulated spray drift from a crop protection program for potatoes." *Integr Environ Assess Manage* 2 (2):105–125.

Associated Press. 2020. "Biden vows to block Alaska mine project if elected." accessed 8/1/2022. https://apnews.com/article/election-2020-anchorage-fish-salmon-alaska-eb3fe94076f884756c4d72cafe71d2f3.

Australia, and New Zealand. 2017. "Weight of evidence, Guidelines for fresh and marine water quality ", accessed 8/16/2021. https://www.waterquality.gov.au/anz-guidelines/resources/key-concepts/weight-of-evidence.

Baker, J. P., D. P. Bernard, S. W. Christensen, M. J. Sale, J. Freda, K. Heltcher, L. Rowe, P. Scanlon, P. Stokes, G. W. Suter, and W. Warren-Hicks. 1990. *Biological Effects of Changes in Surface Water Acid-Base Chemistry*. Washington, DC: National Acid Precipitation Assessment Program.

Barnthouse, L. W., D. L. DeAngelis, R. H. Gardner, R. V. O'Neill, G.W. Suter, and D. S. Vaughan. 1982. *Methodology for Environmental Risk Analysis*. Oak Ridge, TN: Oak Ridge National Laboratory.

Barnthouse, L. W., and G. W. Suter. 1986. *User's manual for ecological risk assessment*. Oak Ridge, TN: Oak Ridge National Laboratory.

Barnthouse, L. W., G. W. Suter, A. E. Rosen, and J. J. Beauchamp. 1987. "Estimating responses of fish populations to toxic contaminants." *Environ Toxicol Chem* 6:811–824.

Barnthouse, L. W. 2013. "Impacts of entrainment and impingement on fish populations: a review of the scientific evidence." *Environ Sci Policy* 31:149–156.

Belanger, S. E., and G. J. Carr. 2020. "Quantifying the precision of ecological risk: Misunderstandings and errors in the methods for assessment factors versus species sensitivity distributions." *Ecotoxicol Environ Saf* 198:110684.

Belanger, S. E., A. Beasley, J. L. Brill, J. Krailler, K. A. Connors, G. J. Carr, M. Embry, M. G. Barron, R. Otter, and A. Kienzler. 2021. "Comparisons of PNEC derivation logic flows under example regulatory schemes and implications for ecoTTC." *Regul Toxicol Pharmacol* 123:104933.

Bellucci, C., G. Hoffman, and S. Cormier. 2010. *An Iterative Approach for Identifying the Causes of Reduced Benthic Macroinvertebrate Diversity in the Willimantic River, Connecticut*. Washington, DC: U.S. Environmental Protection Agency, National Center for Environmental Assessment.

Benbrook, Charles M. 2019. "How did the US EPA and IARC reach diametrically opposed conclusions on the genotoxicity of glyphosate-based herbicides?" *Environ Sci Europe* 31 (1):1–16.

Bennett, Micah G., Sylvia S. Lee, Kate A. Schofield, Caroline Ridley, Susan B. Norton, J. Angus Webb, Susan J. Nichols, Ralph Ogden, and Alexandra Collins. 2018. "Using systematic review and evidence banking to increase uptake and use of aquatic science in decision-making." *Limnol Oceanogr Bull* 27 (4):103–109.

Bernstein, P. L. 1996. *Against the Gods: The Remarkable Story of Risk*. New York: John Wiley & Sons.

Bero, L., A. Anglemyer, H. Vesterinen, and D. Krauth. 2016. "The relationship between study sponsorship, risks of bias, and research outcomes in atrazine exposure studies conducted in non-human animals: Systematic review and meta-analysis." *Environ Internat* 92–93:597–604.

Beyer, W. N., D. J. Audet, G. H. Heinz, D. J. Hoffman, and D. Day. 2000. "Relation of waterfowl poisoning to sediment lead concentrations in the Coeur d'Alene River Basin." *Ecotoxicology* 9 (3):207–218.

Bowler, Diana E, Henning Heldbjerg, Anthony D Fox, Maaike de Jong, and Katrin Böhning-Gaese. 2019. "Long-term declines of European insectivorous bird populations and potential causes." *Conserv Biol* 33 (5):1120–1130.

Breinlinger, S, Tabitha J. Phillips, Brigette N. Haram, Jan Mareš, José A. Martínez Yerena, Pavel Hrouzek, Roman Sobotka, W. Matthew Henderson, Peter Schmieder, Susan M. Williams, James D. Lauderdale, H. Dayton Wilde, Wesley Gerrin, Andreja Kust, John W. Washington, Christoph Wagner, Benedikt Geier, Manuel Liebeke, Heike Enke, Timo H. J. Niedermeyer, and Susan B. Wilde. 2021. "Hunting the eagle killer: A cyanobacterial neurotoxin causes vacuolar myelinopathy." *Science* 371 (6536):eaax9050.

Brock, Theo C. M., Kevin C. Elliott, Anja Gladbach, Caroline Moermond, Jörg Romeis, Thomas-Benjamin Seiler, Keith Solomon, and G. Peter Dohmen. 2021. "Open

Science in regulatory environmental risk assessment." *Integr Environ Assess Manage* 17 (6):1229–1242.

Buchwalter, David B., William H. Clements, and Samuel N. Luoma. 2017. "Modernizing water quality criteria in the United States: a need to expand the definition of acceptable data." *Environ Toxicol Chem* 36 (2):285–291.

Cabral, Reniel B., Darcy Bradley, Juan Mayorga, Whitney Goodell, Alan M. Friedlander, Enric Sala, Christopher Costello, and Steven D Gaines. 2020. "A global network of marine protected areas for food." *Proc Natl Acad Sci* 117 (45):28134–28139.

Cairns, J. Jr., K.L. Dickson, and A.W. Maki, eds. 1978. *Estimating the Hazard of Chemical Substances to Aquatic Life. Vol. STP 657.* Philadelphia, PA: American Society for Testing and Materials.

Cardenuto, João P., and Anderson Rocha. 2022. "Benchmarking scientific image forgery detectors." *Sci Eng Ethics* 28 (4):35.

Carriger, J. F. and M. G. Barron. 2020. "A Bayesian network approach to refining ecological risk assessments: Mercury and the Florida panther (*Puma concolor coryi*)." *Ecol Model* 418:108911.

Catlin, Natasha R., Bradley J. Collins, Scott S. Auerbach, Stephen S. Ferguson, James M. Harnly, Chris Gennings, Suramya Waidyanatha, Glenn E. Rice, Stephanie L. Smith-Roe, and Kristine L. Witt. 2018. "How similar is similar enough? A sufficient similarity case study with *Ginkgo biloba* extract." *Food Chem Toxicol* 118:328–339.

Cedergreen, N. 2014. "Quantifying synergy: a systematic review of mixture toxicity studies within environmental toxicology." *PLoS One* 9 (5):e96580.

CEQ. 1997. *Considering Cumulative Effects Under the National Environmental Policy Act.* Washington, DC: Council on Environmental Quality.

CEQ. 2018. *Fact Sheet: CEQ Report on Environmental Impact Statement Timelines.* Edited by Executive Office of the President Council on Environmental Quality. Washington, DC.

Chambers, R. C. Chief Judge. 2014. Memorandum Opinion and Order, Ohio Valley Environmental Coalition, West Virginia Highlands Conservancy and Sierra Club, V. Elk Run Coal Company, Inc. And Alex Energy, Inc., In CIVIL ACTION NO. 3:12–0785, in The United States District Court for the Southern District of West Virginia. Huntington, WV.

Chapman, P. M. 1990. "The sediment quality triad approach to determining pollution-induced degradation." *Sci Total Environ.* 97/98:815–825.

Chawla, D. S. 2016. "Authors retract study that found pollution near fracking sites." Retraction Watch.

Chibwe, L., I A. Titaley, E. Hoh, and S. L. Massey Simonich. 2017. "Integrated framework for identifying toxic transformation products in complex environmental mixtures." *Environ Sci Technol Lett* 4 (2):32–43.

Chinen, K., and T. Malloy. 2022. "Multi-strategy assessment of different uses of QSAR under REACH analysis of alternatives to advance information transparency." *Int J Environ Res Public Health* 19 (7):4338.

Clayton, A. 2021. *Bernoulli's Fallacy: Statistical Illogic and the Crisis of Modern Science,* Columbia University Press.

Clements, W. H., P. Cadmus, C. J. Kotalik, and B. A. Wolff. 2019. "Context-dependent responses of aquatic insects to metals and metal mixtures: a quantitative analysis summarizing 24 years of stream mesocosm experiments." *Environ Toxicol Chem* 38 (11):2486–2496.

Cormier, S. M., and G. W. Suter. 2008. "A framework for fully integrating environmental assessment." *Environ Manage* 42 (4):543–56.

Cormier, S. M., G. W. Suter, and S. B. Norton. 2010. "Causal characteristics for ecoepidemiology." *Hum Ecol Risk Assess* 16:53–73.

Cormier, S. M., and G. W. Suter. 2013a. "A method for assessing causation of field exposure-response relationships." *Environ Toxicol Chem* 32 (2):272–276.

Cormier S. M., and Suter G. W. 2013b. "A method for deriving water quality benchmarks using field data." *Environ Toxicol Chem* 32:255–262.

Cormier, S. M. and G. W. Suter. 2013c. "Response to Roark et al. 2013." Influence of subsampling and modeling assumptions on the USEPA field-based benchmark for conductivity"." *Integr Environ Assess Manage* 9 (4):677–678.

Cormier, S. M., G. W. Suter, L. Zheng, and G. J. Pond. 2013a. "Assessing causation of the extirpation of stream macroinvertebrates by a mixture of ions." *Environ Toxicol Chem* 32 (2):277–87.

Cormier, S. M., G. W. Suter, and L. Zheng. 2013b. "Derivation of a benchmark for freshwater ionic strength." *Environ Toxicol Chem* 32 (2):263–271.

Cormier, S. M., G. W. Suter, L. Zheng, and G. J. Pond. 2013. "Assessing causation of the extirpation of stream macroinvertebrates by a mixture of ions." *Environ Toxicol Chem* 32 (2):277–287.

Cormier, S., G. Suter, and S. Norton. 2014. "Characteristics and Evidence of Causation." In Ecological Causal Assessment, edited by S Norton, S Cormier and G Suter, 47–64. Boca Raton, FL: CRC Press.

Cormier, S., L. Zheng, G. Suter, and C. Flaherty. 2018. "Assessing background levels of specific conductivity using weight of evidence." *Sci Total Environ* 628–629:1637–1649.

Cormier, Susan M., Glenn W. Suter, Mark B. Fernandez, and Lei Zheng. 2020. "Adequacy of sample size for estimating a value from field observational data." *Ecotoxicol Environ Saf* 203:110992.

Crall, J. 2022. "Glyphosate impairs bee thermoregulation." *Science* 376 (6597):1051–1052. doi:10.1126/science.abq5554.

Cumming, G. 2012. *Understanding the New Statistics: Effect Sizes, Confidence Intervals, and Meta-Analysis.* New York: Routledge.

Cumming, G., and R. Calin-Jageman. 2016. *Introduction to the New Statistics: Estimation, Open Science, and Beyond.* Milton Park, UK: Routledge.

Denworth, L. 2008. *Toxic Truth: A Scientist, a Doctor and the Battle Over Lead.* Boston, MA: Beacon Press.

Department of the Environment. 2013. *Significant Impact Guidelines 1.1- Matters of National Environmental Significance.* Canberra, AU: Australian Government.

Desquilbet, Marion, Laurence Gaume, Manuela Grippa, Régis Céréghino, Jean-François Humbert, Jean-Marc Bonmatin, Pierre-André Cornillon, Dirk Maes, Hans Van Dyck, and David Goulson. 2020. "Comment on "Meta-analysis reveals declines in terrestrial but increases in freshwater insect abundances"." *Science* 370 (6523):eabd8947.

Dreier, David A., Sara I. Rodney, Dwayne R. J. Moore, Shanique L. Grant, Wenlin Chen, Theodore W. Valenti Jr, and Richard A. Brain. 2021. "Integrating exposure and effect distributions with the ecotoxicity risk calculator: Case Studies with crop protection products." *Integr Environ Assess Manage* 17 (2):321–330.

EC. 2003a. *Common Implementation Strategy for the Water Framework Directive: Analysis of Pressures and Impacts.* Luxembourg: European Communities.

EC. 2003b. *Communication from the Commission to the Council, the European Parliament and the European Economic and Social Committee -- A European Environment & Health Strategy.* Brussels, Belgium: Commission of the European Communities.

EC. 2011. Common Implementation Strategy for the Water Framework Directive (2000/60/EC) Guidance Document No. 27 Technical Guidance for Deriving Environmental Quality Standards. Brussels, Belgium: European Commission.

ECETOC. 2012. Categorical approaches, read-across, (Q)SAR. Brussels, Belgium: European Centre for Ecotoxicology and Toxicology of Chemicals.

ECHA. 2010. *Practical Guide 2: How to report weight of evidence.* Helsinki, Finland: European Chemicals Agency.

ECHA. 2016. *New approach methodologies in regulatory science.* Helsinki, Finland: European Chemicals Agency.

Efron, B. 2013. "Bayes' theorem in the 21st century." *Science* 340 (6137):1177–1178.

Efroymson, R. A., W. W. Hargrove, M. J. Peterson, D. S. Jones, W. H. Rose, L. L. Pater, G. W. Suter, and K. A. Reinbold. 2001. *Demonstration of the Military Ecological Risk Assessment Framework (MERAF): Apache Longbow - Hell Fire Missile Test at Yuma Proving Ground.* Oak Ridge, TN: Environmental Science Division, Oak Ridge National Laboratory.

Efroymson, R. A., J. P. Nicolette, and G. W. Suter. 2004. "A framework for net environmental benefit analysis for remediation or restoration of contaminated sites." *Environ Manage* 34:315–331.

EFSA. 2006. "Guidance of the Scientific Committee on a request from EFSA related to uncertainties in dietary exposure assessment." *EFSA J* 438:1–54.

EFSA. 2017. "Guidance on the use of the weight of evidence approach in scientific assessments." *EFSA J* 15 (8):1–69.

EFSA Scientific Committee. 2016. "Guidance to develop specific protection goals options for environmental risk assessment at EFSA, in relation to biodiversity and ecosystem services." *EFSA J* 14 (6):1–50.

Egler, F. E. 1970. *The Way of Science: A Philosophy of Ecology for the Layman.* New York. Hafner Publishing Co.

Ellison, J. 2020. "Data omission in key EPA insecticide study shows need for review of industry analysis." UW News.

EndocrineScience.org. 2022. "Weight of Evidence (WoE)." American Chemistry Council. https://www.endocrinescience.org/glossary/weight-of-evidence-woe/.

Enserink, M. 2017a. "Fishy business: accusations of research fraud roil a tight-knit community of ecologists." *Science* 355 (6331):1255–1257.

Enserink, M. 2017b. "Swedish plastics study fabricated, panel finds: high-profile Science paper had highlighted pollution's threat to larval fish." *Science* 358 (6369):1367.

Enserink, M. 2020. "Study disputes carbon dioxide-fish behavior link: three-year effort fails to replicate alarming findings about effects of ocean acidification." *Science* 367 (6474):128–129.

Enserink, M. 2021. "Does ocean acidification alter fish behavior? Fraud allegations create a sea of doubt." *Science* 372 (6542):560–565.

Enserink, M. 2022. "Star marine ecologist committed misconduct, university says: Finding against Danielle Dixson vindicates whistleblowers who questioned high-profile work on ocean acidification." *Science* 377 (6607):699–700.

Environmental Protection Division. 2019. *Development and Use of Initial Dilution Zones in Effluent Discharge Authorizations*. Vancouver, BC: British Columbia Ministry of Environment and Climate Change Strategy.

Etterson, M. 2020. *User's Manual: SSD Toolbox Version 1.0*. Duluth, MN: U.S. Environmental Protection Agency.

Etterson, M., K. Garber, and E. Odenkirchen. 2017. "Mechanistic modeling of insecticide risks to breeding birds in North American agroecosystems." *PLoS One* 12 (5):e0176998.

EU. 2000. "Directive 2000/60/EC of the European Parliament and of the council of 23 October 2000 establishing a framework for Community action in the field of water policy." *Off J European Commun* 327:1–72.

EU. 2005. *Work Package 1*. Development of the concept of exposure scenarios. General framework of exposure scenarios. Scoping study for technical guidance document on preparing the chemical safety report under REACH, final report. Ispra: European Union, European Chemicals Bureau.

Faber, J. H., S. Marshall, P. J. Van den Brink, and L. Maltby. 2019. "Priorities and opportunities in the application of the ecosystem services concept in risk assessment for chemicals in the environment." *Sci Total Environ* 651:1067–1077.

Fodrie, F. J., K. W. Able, F. Galvez, K. L. Heck, Jr., O. P. Jensen, P. C. López-Duarte, C. W. Martin, R. E. Turner, and A. Whitehead. 2014. "Integrating organismal and population responses of Estuarine Fishes in Macondo spill research." *BioScience* 64 (9):778–788.

Forbes, V. E., P. Calow, V. Grimm, T. I. Hayashi, T. Jager, A. Katholm, A. Palmqvist, R. Pastorok, D. Salvito, R. Sibly, J. Spromberg, J. Stark, and R. A. Stillman. 2011. "Adding value to ecological risk assessment with population modeling." *Human Ecol Risk Assess* 17 (2):287–299.

Fox, D. R., R. A. van Dam, R. Fisher, G. E. Batley, A. R. Tillmanns, J. Thorley, C. J. Schwarz, D. J. Spry, and K. McTavish. 2021. "Recent developments in species sensitivity distribution modeling." *Environ Toxicol Chem* 40 (2):293–308.

Franklin, P., R. Stoffels, J. Clapcott, D. Booker, A. Wagenhoff, and C. Hickey. 2019. *Deriving potential fine sediment attribute thresholds for the National Objectives Framework*. Hamilton, NZ: National Institute of Water & Atmospheric Research Ltd.

Galert, W., and E. Hassold. 2021. "Environmental risk assessment of technical mixtures under the European registration, evaluation, authorisation and restriction of chemicals--a regulatory perspective." *Integr Environ Assess Manage* 17 (3):498–506.

Garten, C. T., Jr., G. W. Suter, and B. G. Blaylock. 1985. *Development and Evaluation of Multispecies Test Protocols for Assessing Chemical Toxicity*. Oak Ridge, TN: Oak Ridge National Laboratory.

Giddings, J. M., T. C. M. Brock, W. Heger, F. Heimbach, S. J. Maund, S. M. Norman, H. T. Ratte, C. Schafers, and M. Streloke. 2002. *Community-Level Aquatic System Studies -- Interpretation Criteria*. Pensacola, FL: SETAC Press.

Giddings, J. M., T. A. Anderson, L. W. Hall, Jr., A. J. Hosmer, R. J. Kendall, R. P. Richards, K. R. Solomon, and W. M. Williams. 2005. *Atrazine in North American Surface Waters: A Probabilistic Risk Assessment*. Pensacola, FL: SETAC Press.

Gigerenzer, G., and U. Hoffrage. 1995. "How to improve Bayesian reasoning without instruction: frequency formats." *Psychol Rev.* 102:684–704.

Gigerenzer, G. 2002. *Calculated Risks*. Simon & Schuster: New York.

Greeley Jr, Mark Stephen, Lynn A. Kszos, Arthur J. Stewart, and John G. Smith. 2011. "Role of toxicity assessment and monitoring in managing the recovery of a wastewater receiving stream." *Environ Manage* 47 (6).

Green, K. C., and J. S. Armstrong. 2015. "Simple versus complex forecasting: the evidence." *J Bus Res* 68 (8):1678–1685.

Hacking, I. 2001. *An Introduction to Probability and Inductive Logic*. Cambridge, UK: Cambridge U. Press.

Hackney, J. D., and W. S. Linn. 1979. "Koch's postulates updated: A potentially useful application to laboratory research and policy analysis in environmental toxicology." *Am Rev Respir Dis* 1119:849–852.

Halbrook, R. S., R. L. Brewer, Jr., and D. A. Buehler. 1999. "Ecological risk assessment of a large river-reservoir: 8. Experimental study of the effects of polychlorinated biphenyls on reproductive success of mink." *Environ Toxicol Chem* 18 (4):649–654.

Hallmann, C. A., R.P.B. Foppen, C.A.M. Van Turnhout, H De Kroon, and E Jongejans. 2014. "Declines in insectivorous birds are associated with high neonicotinoid concentrations." *Nature* 511 (7509):341–343.

Hammons, A. S., ed. 1981. *Methods for Ecological Toxicology: A Critical Review of Laboratory Multispecies Tests*. Ann Arbor, MI: Ann Arbor Science.

Hanson, T. 2021. *Hurricane Lizards and Plastic Squid; The Fraught and Fascinating Biology of Climate Change*. New York: Basic Books.

Health Canada. 2022. "Application of weight of evidence and precaution in risk assessment." Last Modified 11/2/2022. https://www.canada.ca/en/health-canada/services/chemical-substances/fact-sheets/application-weight-of-evidence-precaution-risk-assessments.html.

Hecht, S. A., D. H. Baldwin, C. A. Mebane, T. Hawkes, S. J. Gross, and N. L. Scholz. 2007. An overview of sensory effects on juvenile salmonids exposed to dissolved copper: applying a benchmark concentration approach to evaluate sublethal neurobehavioral toxicity. United States, National Marine Fisheries Service, Northwest Fisheries Science Center.

Hendley, P., C. Holmes, S. Kay, S. J. Maund, K. Z. Travis, and M. Zhang. 2001. "Probabilistic risk assessment of cotton pyrethroids: III. A spatial analysis of the Mississippi, USA, cotton landscape." *Environ Toxicol Chem* 20:669–678.

Herlihy, A. T., and J. C. Sifneos. 2008. "Developing nutrient criteria and classification schemes for wadeable streams in the conterminous US." *J N Am Benthol Soc* 27 (4):932–948.

Higgins J., and J. Thomas. (senior editors). 2021. Cochrane Handbook for Systematic Reviews of Interventions version 6.2. www.training.cochrane.org/handbook: Cochrane, London, UK.

Hill, A. B. 1965. "The environment and disease: association or causation?" *Proc R Soc Med* 58:295–300.

Hines, David E., Rory B. Conolly, and Annie M. Jarabek. 2019. "A quantitative source-to-outcome case study to demonstrate the integration of human health and ecological end points using the aggregate exposure pathway and adverse outcome pathway frameworks." *Environ Sci Technol* 53 (18):11002–11012.

Ho, Clifford K. 2016. "Review of avian mortality studies at concentrating solar power plants." AIP Conference Proceedings.

Housenger, J., K. G. Sappington, M.A. Ruhman, R. Bireley, J. Troiano, and D. Alder. 2016. *Preliminary Pollinator Assessment to Support the Registration Review of Imidacloprid*. Washington, DC: U.S. Environmental Protection Agency, Office of Chemical Safety and Pollution Prevention.

Hughes, C.B., M. Griffiths, and C. Swansborough. 2021. *Framework to Improve the Use of Weight of Evidence in Persistence Assessments*. Oxford, UK: Ricardo Energy & Environment.

Hume, D. 1739. *A Treatice of Human Nature*. Edited by D. F. Norton and M. J. Norton. Oxford, UK: Oxford U. Press.

Hume, D. 1748. *An Inquiry Concerning Human Understanding*. Amherst, NY: Prometheus Books.

IPCC. 2000. *Special Report on Emissions Scenarios*. Cambridge, UK: Cambridge U. Press.

IPCC WGI Technical Support Unit. 2018. IPCC Visual Style Guide for Authors. Geneva, Switzerland: Intergovernmental Panel on Climate Change.

Jackson, S. T. 2021. "Transformational ecology and climate change." *Science* 373 (6559):1085–1086.

Jensen, Olaf P. 2019. "Pesticide impacts through aquatic food webs." *Science* 366 (6465):566–567.

Kahnman, D. 2011. *Thinking Fast and Slow*. New York: Penguin Books.

Kashuba, R., C. Menzie, and L. Martin. 2021. "Risk of cardiovascular disease is driven by different combinations of environmental, medical and behavioral factors: building a conceptual model for cumulative risk assessment." *Human Ecol Risk Assess* 27 (7):1902–1925.

Kass, Robert E., and Adrian E. Raftery. 1995. "Bayes factors." *J Am Stat Assoc* 90 (430):773–795.

Kinnell, P.I.A. 2010. "Event soil loss, runoff and the Universal Soil Loss Equation family of models: a review." *J Hydrol* 385 (1–4):384–397.

Komonen, A., P. Halme, and J. Kotiaho. 2019. "Alarmist by bad design: Strongly popularized unsubstantiated claims undermine credibility of conservation science." *Rethinking Ecol* 4:17–19.

Konemann, H. 1980. "Structure-activity relationships and additivity in fish toxicities of environmental pollutants'". 4: 415–421.

Kosciuch, K., D. Riser-Espinoza, M. Gerringer, and W. Erickson. 2020. "A summary of bird mortality at photovoltaic utility scale solar facilities in the Southwestern U.S." *PLoS One* 15 (4):e0232034. doi: 10.1371/journal.pone.0232034.

Kwok, Kevin WH, Anders Bjorgesæter, Kenneth MY Leung, Gilbert CS Lui, John S Gray, Paul KS Shin, and Paul KS Lam. 2008. "Deriving site-specific sediment quality guidelines for Hong Kong marine environments using field-based species sensitivity distributions." *Environ Toxicol Chem: Int J* 27 (1):226–234.

Lai, A., A. M. Clark, B. I. Escher, M. Fernandez, L. R. McEwen, Z. Tian, Z. Wang, and E. L. Schymanski. 2022. "The next frontier of environmental unknowns: substances of unknown or variable composition, complex reaction products, or biological materials (UVCBs)." Environmental Science & Technology.

Legal Information Institute. 2022. "Weight of Evidence." Cornell Law School. https://www.law.cornell.edu/wex/weight_of_the_evidence.

Leopold, L. B., F. E. Clarke, B. B. Hanshaw, and J. R. Balsley. 1971. *A Procedure for Evaluating Environmental Impact*. Washington, DC: U.S. Geological Survey.

Leung, K. M. Y., J. S. Gray, W. K. Li, G. C. S. Lui, Y. Wang, and P. K. S. Lam. 2005. "Deriving sediment quality guidelines from field-based species sensitivity distributions." *Environ Sci Technol* 39 (14):5148–5156.

Levy, D. 2021. *Maxims for Thinking Analytically*. Monee, IL: Dan Levy.

Lizarraga, L., J. D. Dean, P. Kaiser, S. C. Wesselkamper, J. C. Lambert, and Q. J. Zhao. 2019. "A case study on the application of an expert-driven read-across approach in support of quantitative risk assessment of p, p'-dichlorodiphenyldichloroethane." *Regul Toxicol Pharmacol* 103:301–313.

Lizarraga L., G.W. Suter, J.C. Lambert, G. Patlewicz, J.O. Zhao, J. Dean, and P. Kaiser. 2023. "Advancing the science of a read-across framework for evaluation of data-poor chemicals incorporating systematic and new approach methods." Regul Toxicol Pharmacol 137: 105293.

Lombardo, A., A. Franco, A. Pivato, and A. Barausse. 2015. "Food web modeling of a river ecosystem for risk assessment of down-the-drain chemicals: a case study with AQUATOX." *Sci Total Environ* 508:214–227.

Lonnstedt, O.M., and P. Eklov. 2016. "Environmentally relevant concentrations of microplastic particles influence larval fish ecology." *Science* 352 (6290):1213–1216.

Maehr, D.S., and J.A. Cox. 1995. "Landscape features and panthers in Florida." *Conser Biol* 9 (5):1008–1019.

Mahler, Barbara June, and Peter Van Metre. 2011. Coal-tar-based pavement sealcoat, polycyclic aromatic hydrocarbons (PAHs), and environmental health: US Department of the Interior, US Geological Survey.

Malaeb, Z, J K. Summers, and B. H. Pugesek. 1999. "Using structural equation modeling to investigate relationships among ecological variables." *Environ Ecol Stat* 7:93–111.

March, D., and E. Susser. 2006. "The eco- in eco-epidemiology." *Int J Epidemiol* 35 (6):1379–1383.

Markwiese, James T, Brett Tiller, Randall T Ryti, and Roy Bauer. 2008. "Using artificial burrows to evaluate inhalation risks to burrowing mammals." *Integr Environ Assess Manage* 4 (4):425–430.

Mastrandrea, M D, C B Field, T F Stocker, O Edenhofer, K L Ebi, D J Frame, H Held, E Kriegler, K J Mach, and P R Matschoss. 2010. *Guidance Note for Lead Authors of the IPCC Fifth Assessment Report on Consistent Treatment of Uncertainties*. Geneva, Switzerland: Intergovenmental Program on Climate Change.

Maund, Stephen J., Kim Z. Travis, Paul Hendley, Jeffrey M. Giddings, and Keith R. Solomon. 2001. "Probabilistic risk assessment of cotton pyrethroids: V. Combining landscape-level exposures and ecotoxicological effects data to characterize risks." *Environ Toxicol Chem* 20 (3):687–692.

McFadden, J. 2021. *Life is Simple*. New York: Basic Books.

McLachlan, John A. 1997. "Synergistic effect of environmental estrogens: report withdrawn." *Science* 277 (5325):459–463.

Mebane, C. A., J. P. Sumpter, A. Fairbrother, T. P. Augspurger, T. J. Canfield, W. L. Goodfellow, P. D. Guiney, A. LeHuray, L .Maltby, D. B .Mayfield, M. J. McLaughlin, L. S. Ortego, T. Schlekat, R. P. Scroggins, and T. A. Verslycke. 2019. "Scientific integrity issues in environmental toxicology and chemistry: improving research reproducibility, credibility, and transparency." *Integr Environ Assess Manage* 15 (3):320–344.

Menzie, C., M. H. Henning, J. Cura, K. Finkelstein, J. Gentile, J. Maughan, D. Mitchell, S. Petron, B. Potocki, S. Svirsky, and P. Tyler. 1996. "Special report of the Massachusetts weight-of-evidence workgroup A weight-of-evidence approach for evaluating ecological risks." *Human Ecol Risk Assess* 2 (2):277–304.

Michaels, D. 2008. *Doubt is Their Product*. New York: Oxford U Press.

Moe, S. Jannicke, John F. Carriger, and Miriam Glendell. 2021. "Increased use of Bayesian network models has improved environmental risk assessments." *Integr Environ Assess Manage* 17 (1):53–61.

Moore, Jonathan W, and Daniel E Schindler. 2022. "Getting ahead of climate change for ecological adaptation and resilience." *Science* 376 (6600):1421–1426.

Moretto, A., A. Bachman, A. Boobis, K. R. Solomon, T. P. Pastoor, M. F. Wilks, and M. R. Embry. 2017. "A framework for cumulative risk assessment in the 21st century." *Critical Rev Toxicol* 47 (2):85–97.

Mossler, Max. 2021. "Retraction of flawed MPA study implicates larger problems in MPA science." Sustainable Fisheries. https://sustainablefisheries-uw.org/flawed-mpa-science-retracted/

Munns, W., A.W. Rea, G. Suter, L. R. Martin, L. Blake-Hedges, T. Crk, C. S. Davis, G. Ferreira, S. Jordan, M. Mahoney, and M.G. Barron. 2016. "Ecosystem services as assessment endpoints for ecological risk assessment." *Integr Environ Assess Manage* 12:522–528.

Nabholz, J. V., R. G. Clements, and M. G. Zeeman. 1997. "Information needs for risk assessment in EPA's Office of Pollution Prevention and Toxics." *Ecol Appl* 7 (4):1094–1102.

NAS. 2003. *Cumulative Environmental Effects of Oil and Gas Activities on Alaska's North Slope*. Washington, DC: National Academies Press.

NAS. 2009. *Science and Decisions: Advancing Risk Assessment*. Washington, DC: National Academy of Science.

NCEE. 2010. *Guidelines for Preparing Economic Analyses*. Washington, DC: National Center for Environmental Economics, U.S. Environmental Protection Agency.

NCEE. 2014. *Guidelines for Preparing Economic Analyses*. Washington, DC: National Center for Environmental Economics, U.S. Environmental Protection Agency.

Newman, M. C., D. R. Ownby, L. C. A. Mezin, D. C. Powell, T. R. L. Christensen, S. B. Lerberg, B. Anderson, and T. V. Padma. 2002. "Species sensitivity distributions in ecological risk assessment: distributional assumptions, alternate bootstrap techniques, and estimation of adequate number of species." In Species Sensitivity Distributions in Ecotoxicology, edited by L. Posthuma, G. W. Suter, and T. P. Traas. Boca Raton, FL: Lewis Pub.

NOAA. 1990. *Excavation and Rock Washing Treatment Technology: Net Environmental Benefits Analysis*. Seattle, WA: Hazardous Materials Branch, National Oceanic and Atmospheric Administration.

Norberg-King, T. J., L. W. Ausley, D. T. Burton, W. L. Goodfellow, J. L. Miller, and W. T. Waller. 2005. *Toxicity Reduction and Toxicity Identification Evaluations for Effluents, Ambient Waters, and Other Aqueous Media*. Edited by T. J. Norberg-King, L. W. Ausley, D. T. Burton, W. L. Goodfellow, J. L. Miller and W. T. Waller. Pensacola, FL: SETAC Press.

Norton, S. B., L. Rao, G. W. Suter, and S. Cormier. 2003. "Minimizing cognitive errors in site-specific causal analyses." *Hum Ecol Risk Assess* 9:213–230.

Norton, S.B., and C. L. Schofield. 2018. "Conceptual model diagrams as evidence scaffolds for environmental assessment and management." *Freshwater Sci* 36 (1):231–239.

NRC. 1983. *Risk Assessment in the Federal Government: Managing the Process*. Washington, DC: National Research Council.

NWCC. 2010. Wind Turbine Interactions with Birds, Bats, and their Habitats: A Summary of Research Results and Priority Questions. U.S. National Wind Coordinating Collaborative

O'Neill, R. V., R. H. Gardner, L. W. Barnthouse, G. W. Suter, S. G. Hildebrand, and C. W. Gehrs. 1982. "Ecosystem risk analysis: a new methodology." *Environ Toxicol Chem* 1:167–177.

OECD. 1992. *Report of the OECD Workshop on the Extrapolation of Laboratory Aquatic Toxicity Data to the Real Environment*. Paris: Organization for Economic Cooperation and Development.

OECD. 2004a. *Test No. 218: Sediment-Water Chironomid Toxicity Using Spiked Sediment*. Paris: Organization for Economic Cooperation and Development.

OECD. 2004b. *Test No. 219: Sediment-Water Chironomid Toxicity Using Spiked Water*. Paris: Organization for Economic Cooperation and Development.

OECD. 2021. "The OECD QSAR Toolbox." Paris: Organization for Economic Cooperation and Development. accessed March 14, 2022.

OECD. 2022a. "Adverse Outcome Pathways, Molecular Screening and Toxicogenomics." Paris: Organization for Economic Co-operation and Development, accessed 8/2/2022. https://www.oecd.org/env/ehs/testing/adverse-outcome-pathways-molecular-screening-and-toxicogenomics.htm.

OECD. 2022b. "Hazard assessment." Organization for Economic Cooperation and Development. https://www.oecd.org/chemicalsafety/risk-assessment/hazard-assessment.htm.

OECD. 2022c. *OECD Guidelines for the Testing of Chemicals*. Paris, France: Organization for Economic Cooperation and Development.

Office of Emergency and Remedial Response. 1988. *Guidance for conducting remedial investigations and feasibility studies under CERCLA*. Washington, DC: U.S. Environmental Protection Agency.

Olson, J. R., and C. P. Hawkins. 2012. "Predicting natural base-flow stream water chemistry in the western United States." *Water Resour Res* 48 (2):(W02504).

OPP. 2020. Imidacloprid Proposed Interim Registration Review Decision Case Number 7605. In Docket Number EPA-HQ-OPP-2008-0844 edited by U.S. Environmental Protection Agency Office of Pesticide Programs. www.regulations. gov.

Oppenheimer, M., N. Oreskes, D. Jamieson, K. Brysse, J. O'Reilly, M. Shindell, and M. Wazeck. 2019. Discerning Experts: The Practices of Scientific Assessment for Environmental Policy. University of Chicago Press.

Oreskes, N., and E. M. Conway. 2010. *Merchants of Doubt*. New York: Bloomsbury Press.

Park, R. A., J. S. Clough, and M. Coombs Wellman. 2008. "AQUATOX: Modeling environmental fate and ecological effects in aquatic ecosystems." *Ecol Model* 213 (1):1–15.

Paulik, L. B., C. E. Donald, B. W. Smith, L. G. Tidwell, K. A. Hobbie, L. Kincl, E. N. Haynes, and K. A. Anderson. 2015. "Impact of natural gas extraction on PAH levels in ambient air." *Environ Sci Technol* 49 (8):5203–5210.

Paulik, L. B., C. E. Donald, B. W. Smith, L. G. Tidwell, K. A. Hobbie, Laurel Kincl, E. N. Haynes, and K. A. Anderson. 2016. "Emissions of polycyclic aromatic hydrocarbons from natural gas extraction into air." *Environ Sci Technol* 50 (14):7921–7929.

Pearl, J. and D. Mackenzie. 2018. *The Book of Why: The New Science of Cause and Effect*. New York: Basic Books.

Pearson, K. 1896. "Mathematical Contributions to the Theory of Evolution. III. Regression, Heredity, and Panmixia." *Phil Trans R Soc London* 187:253–318.

Pennisi, E. 2022. "The most unusual birds are also the most at risk." *Science* 377 (6605):458.

Philip, S. Y., S. F. Kew, G.J. van Oldenborgh, F.E.L. Otto, R. Vautard, K. van der Wiel, A.D. King, F.C. Lott, J. Arrighi, and R.P. Singh. 2020. "A protocol for probabilistic extreme event attribution analyses." *Adv Stat Climatol Meteorol Oceanogr* 6: 177–203.

Pinker, S. 2014. *The Sense of Style: The Thinking Person's Guide to Writing in the 21st Century*. New York: Viking.

Pittman, C. 2020. *Cat Tale*. Toronto, Canada: Hanover Square Press.

Pluta, B. 2022. "Freshwater sediment screening benchmarks." U.S. Environmental Protection Agency. accessed 7/31/2022. https://www.epa.gov/risk/freshwater-sediment-screening-benchmarks.

Pollman, C. D., D.G. Rumbold, and D.M. Axelrad, eds. 2020. *Temporal Trends of Mercury in the Everglades, Synthesis and Management Implications*. Geneva: Springer Nature Switzerland AG.

Pond, G. J., M.E. Passmore, F. A. Borsuk, L. Reynolds, and C. J. Rose. 2008. "Downstream effects of mountaintop coal mining: comparing biological conditions using family- and genus-level macroinvertebrate bioassessment tools." *J N Am Benthol Soc* 27 (3):717–737.

Posthuma, L., G. W. Suter, and T. P. Traas, eds. 2002. *Species Sensitivity Distributions for Ecotoxicology*. Boca Raton, FL: Lewis Publishers.

Posthuma, L., J. van Gils, M.C. Zijp, D. van de Meent, and D. de Zwart. 2019. "Species sensitivity distributions for use in environmental protection, assessment, and management of aquatic ecosystems for 12 386 chemicals." *Environ Toxicol Chem* 38 (4):905–917.

Quality Assurance Management Staff. 1994. *Guidance for the data quality objectives process*. Washington, DC: U.S. Environmental Protection Agency.

Raimondo, S, C R Lilavois, and M G Barron. 2016. *Web-based Interspecies Correlation Estimation (Web-ICE) for Acute Toxicity: User Manual Version 3.3*. Gulf Breeze, FL: U.S. Environmental Protection Agency.

Redman, A. D., J. Bietz, J. W. Davis, D. Lyon, E. Maloney, A. Ott, J. C. Otte, F. Palais, J. R. Parsons, and N. Wang. 2021. "Moving persistence assessments into the 21st century: a role for weight-of-evidence and overall persistence." *Integr Environ Assess Manage* 18 (4):868–887.

Richter, L., A. Cordner, and P. Brown. 2021. "Producing ignorance through regulatory structure: the case of per- and polyfluoroalkyl substances (PFAS)." *Sociol Perspect* 64 (4):631–656.

Risk Assessment Forum. 2000. *Supplementary Guidance for Conducting Health Risk Assessment of Chemical Mixtures*. Washington, DC: U.S. Environmental Protection Agency.

Rohr, J. R., and K. A. McCoy. 2010. "Preserving environmental health and scientific credibility: a practical guide to reducing conflicts of interest." *Conserv Lett* 3 (3):143–150.

Roth, N., J. Sandström, and M. F. Wilks. 2020. "A case study applying pathway-oriented thinking to problem formulation for planning a systematic review." *Environ Int* 140:105768.

Rothman, K. J., and S. Greenland. 1998. *Modern Epidemiology, 2nd edition*. Philadelphia, PA: Lippincott, Williams & Wilkins.

Rovida, C., T. Barton-Maclaren, E. Benfenati, F. Caloni, P. C. Chandrasekera, C. Chesné, M. T. D. Cronin, J. De Knecht, D. R. Dietrich, S. E. Escher, S. Fitzpatrick, B. Flannery, M. Herzler, S. Hougaard Bennekou, B. Hubesch, H. Kamp, J. Kisitu, N. Kleinstreuer, S. Kovarich, M. Leist, A. Maertens, K. Nugent, G. Pallocca, M. Pastor, G. Patlewicz, M. Pavan, O. Presgrave, L. Smirnova, M. Schwarz, T. Yamada, and T. Hartung. 2020. "Internationalization of read-across as a validated new approach method (NAM) for regulatory toxicology." *Altex* 37 (4):579–606.

Sala, E, J Mayorga, D Bradley, R B Cabral, T B Atwood, A Auber, W Cheung, C Costello, F Ferretti, and A M Friedlander. 2021. "Protecting the global ocean for biodiversity, food and climate." *Nature* 592 (7854):397–402.

Sample, B. E., and G. W. Suter. 2002. "Screening evaluation of the ecological risks to terrestrial wildlife associated with a coal ash disposal site." *Human Ecol Risk Assess* 8:637–656.

Sample, B. E., Suter G. W., Efroymson R. A., and D. S. Jones. 1998. *A Guide to the ORNL Ecotoxicological Screening Benchmarks: Background, Development, and Application*. Oak Ridge, TN: Oak Ridge National Laboratory.

Sample, B. E., Chris Schlekat, D. J. Spurgeon, C. Menzie, Jon Rauscher, and B. Adams. 2013. "Recommendations to improve wildlife exposure estimation for development of soil screening and cleanup values." *IEAM* 10 (3):372–387.

Sánchez-Bayoa, F., and K. Wyckhuysb. 2019. "Worldwide decline of the entomofauna: a review of its drivers." *Biol Conserv* 232:8–27.

Schipper, A. M., L. Posthuma, D. de Zwart, and M. A. J. Huijbregts. 2014. "Deriving field-based species sensitivity distributions (f-SSDs) from stacked species distribution models (S-SDMs)." *Environ Sci Technol* 48 (24):14464–14471.

Schulz, N. T., and J. F. Berkowitz. 2017. User Guide for Automated Wetland Determination Data Sheets. Wetlands Regulatory Assistance Program, U.S. Army Corps of Engineers.

Science Policy Council. 2000. *Risk Characterization Handbook*. Washington, DC: U.S. Environmental Protection Agency.

Sense About Science. 2013. *Making Sense of Uncertainty: Why Uncertainty Is Part of Science*. London, UK: Sense About Science, archive.senseaboutscience.org.

Sheppard, L., S. McGrew, and R.A. Fenske. 2020. "Flawed analysis of an intentional human dosing study and its impact on chlorpyrifos risk assessments." *Environ Int* 143 (105905):1–8.

Shipley, B. 2000. *Cause and Correlation in Biology: A user's Guide to Path Analysis, Structural Equations, and Causal Inference*. Cambridge, UK: Cambridge U. Press.

Slabe, V. A., J. T. Anderson, B. A. Millsap, J. L. Cooper, A. R. Harmata, M. Restani, R. H. Crandall, B. Bodenstein, P. H. Bloom, and T. Booms. 2022. "Demographic implications of lead poisoning for eagles across North America." *Science* 375 (6582):779–782.

Soskolne, Colin L., Shira Kramer, Juan Pablo Ramos-Bonilla, Daniele Mandrioli, Jennifer Sass, Michael Gochfeld, Carl F. Cranor, Shailesh Advani, and Lisa A. Bero. 2021. "Toolkit for detecting misused epidemiological methods." *Environ Health* 20 (1):90.

Staveley, J. P., J. W. Green, J. Nusz, D. Edwards, K. Henry, M. Kern, A. M. Deines, R. Brain, B. Glenn, and N. Ehresman. 2018. "Variability in nontarget terrestrial plant studies should inform endpoint selection." *Integr Environ Assess Manage* 14 (5):639–648.

Stephan, C. E., D. I. Mount, D. J. Hanson, J. H. Gentile, G. A. Chapman, and W. A. Brungs. 1985. *Guidelines for Deriving Numeric National Water Quality Criteria for the Protection of Aquatic Organisms and Their Uses.* Washington, DC: U.S. Environmental Protection Agency.

Stokstad, E. 2021. "Mysterious eagle killer identified." *Science* 371 (6536):1298–1298.

Suter, G. W. 1983. "Multispecies tests for environmental toxicology." *Environ Int* 9:157–160.

Suter, G. W., D. S. Vaughan, and R. H. Gardner. 1983. "Risk assessment by analysis of extrapolation error, a demonstration for effects of pollutants on fish." *Environ Toxicol Chem* 2:369–378.

Suter, G. W. 1989. "Ecological endpoints." In *Ecological Assessment of Hazardous Waste Sites: A Field and Laboratory Reference Document. EPA 600/3-89/013,* edited by W. Warren-Hicks, B. R. Parkhurst and S.S. Baker, Jr., Corvallis, Oregon: Corvallis Environmental Research Laboratory.

Suter, G. W. 1990. "Use of biomarkers in ecological risk assessment." In *Biomarkers of Environmental Contamination,* edited by J. F. McCarthy and L. L. Shugart, 419–426. Ann Arbor, MI: Lewis Publishers.

Suter, G. W. 1993a. "A critique of ecosystem health concepts and indices." *Environ Toxicol Chem* 12:1533–1539.

Suter, G. W. 1993b. *Ecological Risk Assessment.* Boca Raton, FL: Lewis Publishers.

Suter, G. W. 1996. "Abuse of hypothesis testing statistics in ecological risk assessment." *Human Ecol Risk Assess* 2:331–349.

Suter, G. W. 1997. "Integration of human health and ecological risk assessment." *Environ Health Persp* 105:1282–1283.

Suter, G. W. 1998. "Comments on the interpretation of distributions in "Overview of recent developments in ecological risk assessment." *Risk Anal* 18:3–4.

Suter, G. W. 1999. "Developing conceptual models for complex ecological risk assessments." *Human Ecol Risk Assess* 5 (1):375–396.

Suter, G. W., R. A. Efroymson, B. E. Sample, and D. S. Jones. 2000. *Ecological Risk Assessment for Contaminated Sites.* Boca Raton, FL: Lewis Publishers.

Suter, G. W., K. A. Reinbold, W. H. Rose, and M. K. Chawla. 2002. *Military Ecological Risk Assessment Framework (MERAF) for Assessment of Risks of Military Training and Testing to Natural Resources.* Oak Ridge, TN: Environmental Sciences Division, Oak Ridge National Laboratory.

Suter, G. W., T. Vermier, W. R. Munns, Jr., and J. Sekizawa. 2003. "Framework for the integration of health and ecological risk assessment." *Human Ecol Risk Assess* 9:281–302.

Suter, G. W. 2004. "Bottom-up and top-down integration of human and ecological risk assessment." *Toxicol. Environ. Health, Part A* 67:779–790.

Suter, G. W., S. B. Norton, and A. Fairbrother. 2005. "Individuals versus organisms versus populations in the definition of ecological assessment endpoints." *Integr Environ Assess Manage* 1:397–400.

Suter, G. W. 2007. *Ecological Risk Assessment, 2nd edition.* Boca Raton, FL: CRC Press.

Suter, G. W. and S. M. Cormier. 2008. "What is meant by risk-based environmental quality criteria?" *Integr Environ Assess Manage* 4:486–489.

Suter, G. W., and T. P. O'Farrell. 2009. Analysis of the causes of a decline in the San Joaquin kit fox population on the Elk Hills Naval Petroleum Reserve #1, California. Cincinnati, OH: National Center for Environmental Assessment, Office of Research and Development, U.S. EPA.

Suter, G. W. and S. M. Cormier. 2012. "Two roles for environmental assessors: Technical Consultant and Advisor." *Human Ecol Risk Assess* 18 (6):1153–1155.

Suter, G. W. and S. M. Cormier. 2013. "A method for assessing the potential for confounding applied to ionic strength in central Appalachian streams." *Environ Toxicol Chem* 32 (2):288–295.

Suter, G. W., and S. Cormier. 2014. "The problem of biased data and potential solutions for health and environmental assessments." *Human Ecol Risk Assess: Int J* 21:1736–1752.

Suter, G. W. and T. P. O'Farrell. 2015. "Applying CADDIS to a terrestrial case: San Joaquin kit foxes on an oil field." In *Ecological Causal Assessment*, edited by S. Norton, S. Cormier and G. Suter, 397–434. Boca Raton, FL: CRC Press.

Suter, G. W., and S. M. Cormier. 2015a. "Bias in the development of health and ecological assessments and potential solutions." *Human Ecol Risk Assess* 22:99–115.

Suter, G. W., and S. M. Cormier. 2015b. "Why care about aquatic insects: uses, benefits, and services." *Integr Environ Assess Manage* 11 (2):188–194.

Suter, G. W., S. M. Cormier, and M. G. Barron. 2017. "A weight of evidence framework for environmental assessments: inferring quantities." *Integr Environ Assess Manage* 13:1045–1051.

Suter, G. W. 2018. "Specifying the dimensions of aquatic life benchmark values in clear, complete, and justified problem formulations." *Integr Environ Assess Manage* 14:631–638.

Suter, G. W. 2019. "Statistics cannot decide how much to protect the environment." *Integr Environ Assess Manage* 15 (4):495–496.

Suter, G. W., J. Nichols, E. Lavoie, and S. M. Cormier. 2020a. "The origins of classic WoE and SR, supplemental material for: systematic review and weight of evidence are integral to ecological and human health assessments: they need an integrated framework." *IEAM* 16 (5):718–728.

Suter, G., E. Lavoie, J. D. Nichols, and S. Cormier. 2020b. "Systematic review and weight of evidence are integral to ecological and human health assessments: they need an integrated framework." *IEAM* 16:718–728.

Suter, G. W. and L. Lizarraga. 2022. "Clearly weighing the evidence in read-across can improve assessments of data-poor chemicals." *Regul Toxicol Pharmacol* 129: article 105111, doi.org/10.1016/j.yrtph.2021.105111.

Tallamy, D. W., and W. G. Shriver. 2021. "Are declines in insects and insectivorous birds related?" *Ornithol Appl* 123 (1). doi: 10.1093/ornithapp/duaa059.

Then, C., K. Kawall, and N. Valenzuela. 2020. "Spatiotemporal controllability and environmental risk assessment of genetically engineered gene drive organisms from the perspective of European union genetically modified organism regulation." *Integr Environ Assess Manage* 16 (5):555–568.

Tian, Z., H. Zhao, K. T. Peter, M. Gonzalez, J. Wetzel, C. Wu, X. Hu, J. Prat, E. Mudrock, R. Hettinger, A. E. Cortina, R. G. Biswas, F. V. C. Kock, R. Soong, A. Jenne, B. Du, F. Hou, H. He, R. Lundeen, A. Gilbreath, R. Sutton, N. L. Scholz, J. W. Davis, M. C. Dodd, A. Simpson, J. K. McIntyre, and E. P. Kolodziej. 2021. "A ubiquitous tire rubber-derived chemical induces acute mortality in coho salmon." *Science* 371 (6525):185–189.

Tidwell, L. G., S. E. Allan, S. G. O'Connell, K. A. Hobbie, B. W. Smith, and K. A. Anderson. 2015. "Polycyclic aromatic hydrocarbon (PAH) and oxygenated PAH (OPAH) air–water exchange during the deepwater horizon oil spill." *Environ Sci Technol* 49 (1):141–149.

Tidwell, L. G., S. E. Allan, S. G. O'Connell, K. A. Hobbie, B. W. Smith, and K. A. Anderson. 2016. "PAH and OPAH Flux during the Deepwater Horizon Incident." *Environ Sci Technol* 50 (14):7489–7497.

Traas, TP., D. van de Meent, L. Posthuma, T. Hamers, BJ Kater, D. de Zwart, and T. Aldenberg. 2002. "The potentially affected fraction as a measure of ecological risk." In *Species Sensitivity Distributions in Ecotoxicology*, edited by L. Posthuma, G Suter and T. Traas, 587. Boca Raton, FL: Lewis Publishers.

Tuart, L. W. 1988. *Hazard Evaluation Division Technical Guidance Document: Aquatic Mesocosm Tests to Support Pesticide Registration*. Washington, DC: U.S. Environmental Protection Agency.

Tuart, L. W., and A. F. Maciorowski. 1997. "Information needs for pesticide registration in the United States." *Ecol Appl* 7 (4):1086–1093.

USC. 2016. Frank R. *Lautenberg Chemical Safety for the 21st Century Act*. Public Law 114–182. Washington, DC: United States Congress.

USDHEW. 1964. *Smoking and Health: Report of the Advisory Committee to the Surgeon General*. Washington, DC: U.S. Department of Health, Education and Welfare.

USEPA. 1985. *Guidelines for Deriving Numeric National Water Quality Criteria for the Protection of Aquatic Organisms and Their Uses*. Washington, DC: U.S. Environmental Protection Agency.

USEPA. 1989. *Use of Starling Nest Boxes for Field Reproductive Studies*. Corvallis, OR: Office of Research and Development, U.S. Environmental Protection Agency.

USEPA. 1992. *Framework for Ecological Risk Assessment*. Washington, D.C.: Risk Assessment Forum, U.S. Environmental Protection Agency.

USEPA. 1994. *Using Toxicity Tests in Ecological Risk Assessment*. Washington, DC: Office of Emergency and Remedial Response, U.S. Environmental Protection Agency.

USEPA. 1998. *Guidelines for Ecological Risk Assessment*. Washington, DC: Risk Assessment Forum, U.S. Environmental Protection Agency.

USEPA. 1999. *Toxicity Reduction Evaluation Guidance for Municipal Wastewater Treatment Plants*. Washington, DC: Office of Wastewater Management, U.S. Environmental Protection Agency.

USEPA. 2000a. *Choosing a Percentile of Acute Dietary Exposure as a Threshold for Regulatory Concern*. Washington, DC: U.S. Environmental Protection Agency.

USEPA. 2000b. *Stressor Identification Guidance Document*. Washington, DC: Office of Water, Office of Research and Development, U.S. Environmental Protection Agency.

USEPA. 2001. *Risk Assessment Guidance for Superfund. Vol. III, Part A. Process for Conducting Probabilistic Risk Assessment (RAGS 3A)*. Washington, DC: U.S. Environmental Protection Agency.

USEPA. 2002a. *Methods for measuring the acute toxicity of effluents to freshwater and marine organisms, Fifth edition*. Washington, DC: Office of Water, U.S. Environmental Protection Agency.

USEPA. 2002b. *Procedures for the Derivation of Equilibrium Partitioning Sediment Benchmarks (ESBs) for the Protection of Benthic Organisms: Dieldrin*. Washington, DC: Office of Water, U.S. Environmental Protection Agency.

USEPA. 2002c. *Procedures for the Derivation of Equilibrium Partitioning Sediment Benchmarks (ESBs) for the Protection of Benthic Organisms: Endrin*. Washington, DC: Office of Water, U.S. Environmental Protection Agency.

USEPA. 2002d. *Procedures for the Derivation of Equilibrium Partitioning Sediment Benchmarks (ESBs) for the Protection of Benthic Organisms: Metal Mixtures*. Washington, DC: Office of Water, U.S. Environmental Protection Agency.

USEPA. 2002e. *Procedures for the Derivation of Equilibrium Partitioning Sediment Benchmarks (ESBs) for the Protection of Benthic Organisms: PAH Mixtures*. Washington, DC: Office of Water, U.S. Environmental Protection Agency.

USEPA. 2002f. *Short-Term Methods for Estimating the Chronic Toxicity of Effluents and Receiving Waters to Freshwater Organisms -- Fourth Edition*. Washington, DC: U.S. Environmental Protection Agency.

USEPA. 2002g. *Short-Term Methods for Estimating the Chronic Toxicity of Effluents and Receiving Waters to Marine and Estuarine Organisms -- Third Edition*. Washington, DC: U.S. Environmental Protection Agency.

USEPA. 2003a. *Developing Relative Potency Factors for Pesticide Mixtures: Biostatistical Analysis of Joint Dose-Response*. Cincinnati, OH: Office of Research and Development. U.S. Environmental Protection Agency.

USEPA. 2003b. *Framework for Cumulative Risk Assessment*. Washington, DC: U.S. Environmental Protection Agency.

USEPA. 2003c. *Generic Ecological Assessment Endpoints (GEAEs) for Ecological Risk Assessment*. Washington, DC: Risk Assessment Forum, U.S. Environmental Protection Agency.

USEPA. 2005a. *Guidance for Developing Ecological Soil Screening Levels (ECO-SSLs)*. Washington, DC: Office of Solid Waste and Emergency Response, U.S. Environmental Protection Agency.

USEPA. 2005b. *Guidelines for Carcinogen Risk Assessment*. Washington, DC: Risk Assessment Forum, U.S. Environmental Protection Agency.

USEPA. 2006. *Framework for Developing Suspended and Bedded Sediments (SABS) Water Quality Criteria*. Washington, DC: U.S. Environmental Protection Agency.

USEPA. 2007a. *Aquatic Life Ambient Freshwater Quality Criteria - Copper*. Washington, DC: U.S. Environmental Protection Agency.

USEPA. 2007b. *Concepts, Methods and Data Sources for Cumulative Health Risk Assessment of Multiple Chemicals, Exposures and Effects: A Resource Document*. Cincinnati, OH: U.S. Environmental Protection Agency, Office of Research and Development, National Center for Environmental Assessment, U.S. Environmental Protection Agency.

USEPA. 2007c. Considerations for developing a dosimetry-based cumulative risk assessment approach for mixtures of environmental contaminants. EPA/600/R-07/064. Office of Research and Development, National Center for Environmental Assessment, U.S. Environmental Protection Agency.

USEPA. 2008. *Child-Specific Exposure Factors Handbook*. Washington, DC: U.S. Environmental Protection Agency.

USEPA. 2009. *Guidance on the Development, Evaluation, and Application of Environmental Models*. Washington, DC: Council for Regulatory Environmental Modeling, U.S. Environmental Protection Agency.

USEPA. 2010. *Integrating Ecological Assessment and Decision-Making at EPA: A Path Forward. Results of a Colloquium in Response to Science Advisory Board and National Research Council Recommendations*. Washington, DC: Risk Assessment Forum, U.S. Environmental Protection Agency.

USEPA. 2011a. *Exposure Factors Handbook 2011 Edition*. Washington, DC: U.S. Environmental Protection Agency.

USEPA. 2011b. *A Field-based Aquatic Life Benchmark for Conductivity in Central Appalachian Streams*. Washington, DC: Office of Research and Development, National Center for Environmental Assessment, U.S. Environmental Protection Agency.

USEPA. 2013. *Aquatic Life Ambient Water Quality Criteria for Ammonia – Freshwater*. Washington, DC: Office of Water, U.S. Environmental Protection Agency.

USEPA. 2014a. *An Assessment of Potential Mining Impacts on Salmon Ecosystems of Bristol Bay, Alaska. EPA 910-R-001A*. Seattle, WA: U.S. Environmental Protection Agency.

USEPA. 2014b. *Executive Summary of the Final Report, An Assessment of Potential Mining Impacts on Salmon Ecosystems of Bristol Bay, Alaska*. Seattle, WA: Region 10, U.S. Environmental Protection Agency.

USEPA. 2014c. *Framework for Human Health Risk Assessment to Inform Decision Making*. Washington, DC: U.S. Environmental Protection Agency.

USEPA. 2014d. *Policy Assessment for the Review of the Ozone National Ambient Air Quality Standards*. Washington, DC: U.S. Environmental Protection Agency.

USEPA. 2014e. *Welfare Risk and Exposure Assessment for Ozone*. Washington, DC: U.S. Environmental Protection Agency.

USEPA. 2016a. *Aquatic Life Ambient Water Quality Criterion for Selenium – Freshwater*. Washington, DC: Office of Water, U.S. Environmental Protection Agency.

USEPA. 2016b. *Ecosystem Services as Assessment Endpoints in Ecological Risk Assessment*. Washington, DC: U.S. Environmental Protection Agency.

USEPA. 2016c. *Generic Ecological Assessment Endpoints (GEAEs) for Ecological Risk Assessment: Second Edition with Generic Ecosystem Services Endpoints Added*. Washington, DC: Risk Assessment Forum, U.S. Environmental Protection Agency.

USEPA. 2016d. *Office of Pesticide Programs' Framework for Incorporating Human Epidemiology and Incident Data in Risk Assessments of Pesticides*. Washington, DC: U.S. Environmental Protection Agency.

USEPA. 2016e. *Weight of Evidence in Ecological Assessment*. Washington, DC: U.S. Environmental Protection Agency, Risk Assessment Forum.

USEPA. 2018a. *Application of Systematic Review in TSCA Risk Evaluations*. Washington, DC: U.S. Environmental Protection Agency, Office of Chemical Safety and Pollution Prevention.

USEPA. 2018b. *Strategic Plan to Promote the Development and Implementation of Alternative Test Methods Within the TSCA Program*. Washington, DC: U.S. Environmental Protection Agency, Office of Chemical Safety and Pollution Prevention.

USEPA. 2019a. "Clean Water Act Hazardous Substances Spill Prevention." Federal Register 84 FR 46100:46100–46135.

USEPA. 2019b. *Guidelines for Human Exposure Assessment*. Washington, DC: Guidelines for Human Exposure Assessment.

USEPA. 2019c. *IRIS Assessment Plan for Methylmercury (Scoping and Problem Formulation Materials)*. Washington, DC: U.S. Environmental Protection Agency.

USEPA. 2020a. "Ecological Committee on FIFRA Risk Assessment Methods (ECOFRAM)." U.S. Environmental Protection Agency. accessed 6/24/2021. https://www.epa.gov/pesticide-science-and-assessing-pesticide-risks/ecological-committee-fifra-risk-assessment-methods.

USEPA. 2020b. *Policy Assessment for the Review of the National Ambient Air Quality Standards for Particulate Matter.* Washington, DC: Office of Air Quality Planning and Standards, U.S. Environmental Protection Agency.

USEPA. 2021a. *Ambient Water Quality Criteria to Address Nutrient Pollution in Lakes and Reservoirs. EPA-822-R-21-005.* Washington, DC: Office of Water, U.S. Environmental Protection Agency.

USEPA. 2021b. "AQUATOX: Linking water quality and aquatic life." accessed 8/3/2022. https://www.epa.gov/ceam/aquatox.

USEPA. 2021c. "Biological Evaluation Chapters for Chlorpyrifos ESA Assessment." Office of Pesticide Programs, U.S. Environmental Protection Agency, accessed 04/12/2022, https://www.epa.gov/endangered-species/biological-evaluation-chapters-chlorpyrifos-esa-assessment.

USEPA. 2022a. "Biological Evaluation Chapters for Malathion ESA Assessment, Attachment 1–5. Method For Deriving Species Sensitivity Distributions for Use in Pesticide Effects Determinations for Listed Species." accessed 9/15/2022. https://www.epa.gov/endangered-species/biological-evaluation-chapters-malathion-esa-assessment.

USEPA. 2022b. *Metals Cooperative Research and Development Agreement (CRADA) Phase I Report: Development of an Overarching Bioavailability Modeling Approach to Support US EPA's Aquatic Life Water Quality Criteria for Metals.* Washington, DC: Office of Water, U.S. Environmental Protection Agency.

USEPA. 2022c. "Provisional Peer-Reviewed Toxicity Values (PPRTVs)." U.S. Environmental Protection Agency, Last Modified 6/22/2022, accessed 10/7/2022. https://www.epa.gov/pprtv#:~:text=A%20Provisional%20Peer%2DReviewed%20Toxicity, and%20guidance%20for%20value%20derivation.

USEPA. 2022d. "Regional Screening Levels (RSLs)." U.S. Environmental Protection Agency.accessed3/8/2022.https://www.epa.gov/risk/regional-screening-levels-rsls.

USEPA. 2022e. "Risk Management." U.S. Environmental Protection Agency. accessed 5/31/2022. https://www.epa.gov/risk/risk-management#tab-2.

USEPA. 2022f. "Support Center for Regulatory Atmospheric Modeling (SCRAM)." U.S. Environmental Protection Agency. accessed 8/2/2022. https://www.epa.gov/scram.

USEPA Scientific Integrity Office. 2012. U.S. EPA Scientific Integrity Policy. Washington, D.C.: U.S. Environmental Protection Agency.

USEPA/PMRA/CDPR. 2014. *Guidance for Assessing Pesticide Risks to Bees.* Washington, DC: Office of Pesticide Programs, United States Environmental Protection Agency; Health Canada Pest Management Regulatory Agency; California Department of Pesticide Regulation.

Van Straalen, N. M. 2002. "Theory of ecological risk assessment based on species sensitivity distributions." In *Species Sensitivity Distributions in Ecotoxicology*, edited by L. Posthuma, G. W. Suter, and T. Traas, 37–48. Boca Raton, FL: Lewis Pub.

vanKlink, R., D. Bowler, K. Konstantin B. Gongalsky, A. Swengel, and J. Chase. 2020. "Response to Comment on "Meta-analysis reveals declines in terrestrial but increases in freshwater insect abundances"." *Science* 370 (6523):eabe0760.

vanKlink, R., D. Bowler, K. Konstantin B. Gongalsky, A. Swengel, A. Gentile, and J. Chase. 2020. "Meta-analysis reveals declines in terrestrial but increases in freshwater insect abundances." *Science* 368 (6489):417–420.

Veith, G. D., D. J. Call, and L. T. Brook. 1983. "Structure-toxicity relationships for fathead minnow, *Pimephales promelas*: Narcotic industrial chemicals." *Can J Fish Aquat Sci* 40:743–748.

Veith, G. D., and P. Kosian. 1983. "Estimating bioconcentration potential from octanol/water partitioning coefficients." In *PCBs in the Great Lakes*, edited by D. R. Mackay, S. Patterson, S. Eisenreich and M. Simmons. Ann Arbor, MI: Ann Arbor Press.

Vermier, T., W. Munns, J. Sekizawa, G. Suter, and G. Van der Kraak. 2007. "An assessment of integrated risk assessment." *Human Ecol Risk Assess* 13:339–354.

Wagner, W.E., and S.C. Gold. 2022. "Legal obstacles to toxic chemical research." *Science* 375 (6577):138–141.

Walls, S. J., D. S. Jones, A. R. Stojak, and N. E. Carriker. 2015. "Ecological risk assessment for residual coal fly ash at Watts Bar Reservoir, Tennessee: site setting and problem formulation." *Integr Environ Assess Manage* 11 (1):32–42.

Warne, M. S. J., R. A. van Dam, G. E. Batley, and J. L. Stauber. 2017. "Response to Buchwalter et al. Further considerations for modernizing water quality criteria in the United States and elsewhere." *Environ Toxicol Chem* 36 (6):1422–1424.

Wasserstein, R. L., and N. A. Lazar. 2016. "The ASA's statement on p-values: context, process, and purpose." *Am Stat* 70 (2):129–133.

Weidenmüller, Anja, Andrea Meltzer, Stefanie Neupert, Alica Schwarz, and Christoph Kleineidam. 2022. "Glyphosate impairs collective thermoregulation in bumblebees." *Science* 376 (6597):1122–1126.

WHO. 2001. *Report on Integrated Risk Assessment, WHO/IPCS/IRA/01/12*. Geneva, Switzerland: World Health Organization.

WHO. 2008. Uncertainty and Data Quality in Exposure Assessment. The International Programme on Chemical Safety, World Health Organization.

Wigger, H., D. Kawecki, B. Nowack, and V. Adam. 2020. "Systematic consideration of parameter uncertainty and variability in probabilistic species sensitivity distributions." *Integr Environ Assess Manage* 16 (2):211–222.

Wilson, E. O. 1998. *Consilience: The Unity of Knowledge*. New York: A. A. Knopf.

Woodman, J. N., and E. B. Cowling. 1987. "Airborne chemicals and forest health." *Environ Sci Technol* 21:120–126.

Yamamuro, M., T. Komuro, H. Kamiya, T. Kato, H. Hasegawa, and Y. Kameda. 2019. "Neonicotinoids disrupt aquatic food webs and decrease fishery yields." *Science* 366 (6465):620–623.

Yerushalmy, J. and C. E. Palmer. 1959. "On the methodology of investigations of etiologic factors in chronic disease." *J Chronic Disease* 10 (1):27–40.

Index

Note: Bold page numbers refer to tables and *italic* page numbers refer to figures.